CLIMATE

All is well,
All will be well

Jeremy Nieboer

www.brugesgroup.com

Published in 2021 by The Bruges Group, 246 Linen Hall, 162-168 Regent Street, London W1B 5TB

Follow us on Twitter @brugesgroup, LinkedIn @brugesgroup, GETTR @brugesgroup

Facebook The Bruges Group, Instagram brugesgroup, YouTube brugesgroup

ABOUT THE AUTHOR

JEREMY NIEBOER was educated at Harrow School and Oriel College Oxford. After a period practising as a member of the Bar in Kings Bench Walk he was later admitted as a solicitor becoming a partner in two City law firms. He specialised in corporate work, including mergers and acquisitions, capital market public offerings, private equity transactions and commercial law. He still acts for a few long-standing clients. His first encounter with any challenge to the accepted doctrine of 'global warming' came through his contact with Christopher Booker, whom Jeremy first met when acting as lead solicitor on the application by Lord Rees-Mogg to restrain ratification of the Maastricht Treaty. Christopher himself published his essential work "The Real Global Warming Disaster" in 2009. Just at the time of its publication, there was a public meeting in Church House addressed by Professor Plimer in which he succinctly set out the fundamental scientific flaws of alleged CO_2 driven global warming. It was this that set Jeremy on a path of enquiry and research. Jeremy has since spoken at numerous meetings as to the want of any tenable scientific basis for the vast proposed expenditure on the folly of de-carbonisation. He published his first booklet on climate alarmism with the Bruges Group in 2010, "A Lesson in Democracy", and has been a lead speaker at public meetings and debates on the subject.

CONTENTS

FIGURES

FOREWORD

"Let him not be taught science, let him discover it. If ever you substitute authority for reason he will cease to reason; he will be a mere plaything of other people's thoughts".

This precept of Jean Jacques Rousseau of 1762 prefigures the words of Richard Feynman "I would rather have questions that can't be answered than answers that can't be questioned". Realisation arising from enquiry and discovery is enlightening. Belief in authority stifles reason and enquiry. It is as Einstein said "the enemy of truth".

The cult of 'global warming' depends on constant and ever more dire affirmations of great falsehoods. At the root of the cult is the 'enemy' – humans of the developed nations - and the 'big lie' – that humans by their emissions of CO_2 have caused dangerous global warming.

What makes the cult so pernicious is that it traduces our natural wish to look after our beautiful planet - to do what we can to diminish our claims on its resources, not to inflict harm and to restore what we have despoiled. The movements to protect marine life and the oceans from pollution, to replace plastic where we can with degradable material, to limit pesticides and restrict emissions of sulphur dioxide, carbon monoxide and other dangerous gasses – all these have come about gradually over the past 40 years. They are part of our common concern.

It is this goodwill and natural disposition to care for the Earth that the cult of global warming has preyed upon, abusing science with false predictions of catastrophe so as to bring about the overturning of the energy basis of western economies.

Because it is a cult it suppresses contrary opinion, vilifies non-believers and admits no doubt. Its dogma cannot be overthrown by reason and evidence. It is a cult of unreason and it is the cult itself that must be exposed in order to allow those who will to listen once more to the voice of doubt and find the inspiration to enquire for themselves. It is chilling, but nevertheless apt, to consider the words one who so manipulated the truth as to drug a nation and subsume its peoples' reason.

> "In the big lie there is always a certain force of credibility; because the broad masses of a nation are always more easily corrupted in the deeper strata of their emotional nature than consciously or voluntarily; and thus ..they more readily fall victims to the big lie than the small lie, since they themselves often tell small lies in little matters but would be ashamed to resort to large-scale falsehoods. It would never come into their heads to fabricate colossal untruths and they would not believe others could have the impudence to distort the truth so infamously. Even though the facts which prove this to be so may be brought clearly to their minds, they will still doubt and waver and will continue to think there may be some other explanation. For the grossly impudent lie always leaves traces behind it, even after it has been nailed down, a fact which is known to all expert liars in this world and to all who conspire together in the art of lying".

However there is one great gift and assurance that we in this generation possess. That is the fund of observations, research and knowledge that is afforded to us by the internet. It is possible for each of us to draw out of this store scientific papers of deep scholarship which may not otherwise have been published such is the grip of the dogma on means of expression. Within one minute we may see the entire 43 year satellite record of the Earth's temperature. All that is needed is an awakening of doubt and the resurrection of the spirit of enquiry.

Thus it has been possible for this little booklet to be produced by an elderly lawyer with no scientific experience who, on setting on a path of enquiry in 2008, discovered the origins and purpose of the big lie and the wickedness of those who, knowing it to be false, have thereby sought to bring about the overthrow of the basis of a prosperity and welfare such as humanity has never before enjoyed.

OUR CLIMATE. ALL IS WELL. ALL WILL BE WELL

ABSTRACT

- CO_2 is essential to life on Earth, terrestrial and marine. Increases in CO_2 are beneficial since they lead to surplus food, expansion of vegetation and forests. CO_2 regulates the solar energy cycle.

- The addition of more CO_2 to the atmosphere will only result in negligible increases in its greenhouse effect and no further appreciable warming.

- The IPCC has not proved that the increase in global temperature since 1850 is caused by a rise in CO_2 density and that it is not due to heightened solar activity as we emerge from the Little Ice Age. There has been no warming trend since 1998.

- There must be a necessary correlation between the cause of a phenomenon and its effect. There is close correlation of temperature with solar activity. There is no causal correlation of rise in temperature and CO_2 density in any age.

- Climate environmentalism uses science as dogma for political ends. It does so by promoting fear and guilt. The IPCC is a political bureaucracy. It uses borrowed prestige and authority to displace rational enquiry. Consensus cannot prevail over evidence nor does any consensus of climate scientists exist. Scepticism is the essence of science.

- The IPCC abuses the scientific method. It was founded to promote belief in dangerous global warming and procure the overthrow of the energy base of modern economies. Its models are disproved by satellite observational evidence. It knowingly uses false graphs and distortions of data. It fabricates data to conform to its dogma.

- The 2021 Summary for Policymakers knowingly misrepresents the global temperature record. For almost all of the last 10,000 years average temperature has been higher than the past 1,500 years. The IPCC graphs knowingly, with intent to deceive, exclude centuries of pronounced warming and cooling that are inconsistent with its hypothesis.

- Extreme weather events, changes in sea levels, coral growth and extinctions are not indicators of changes in climate much less global warming induced by human emissions.

- The hypothesis that CO_2 is causing dangerous global warming and climate change is false. Achieving Net Zero emissions will have no effect on global temperature.

Summary of Contents

Part 1 of this analysis is limited to just three fundamental matters that displace the predictions of catastrophe in the 'reports' of the UN Intergovernmental Panel on Climate Change (**IPCC**).

1. **No impact of increased CO_2 on global temperature**
   The lack of sensitivity of climate sensitivity of atmospheric CO2 above 250 parts per million (ppm) is such that if the current content of 417 ppm were doubled there would be negligible impact on the radiation balance and accordingly no rise in temperature.

2. **No global temperature abnormality**
   The satellite and balloon radiosonde records of temperature variation over the period since such measuring processes became available (1979) reveal no abnormal rise in temperature.

3. **No correlation of temperature and CO_2 density**
   There is no causative correlation between temperature variation and CO_2 intensity neither in modern times nor over geological time.

Any one of these invalidates the fundament of the IPCC that human emissions of CO_2 are causing dangerous global warming which can only be arrested by dismantling the entire hydrocarbon basis of energy supply.

Part 2 addresses the origins of modern climate environmentalism, the founding purpose of the IPCC, the failure of the IPCC to apply the discipline of the scientific method, its use of misleading and distorted data and its publication of falsified graphs.

It concludes with a searching examination of the latest Summary for Policymakers (August 2021). That document is the basis of the determinations of the Climate Change Conference in Glasgow (31st October to 12th November 2021) chaired by the UK.

The **Appendix** addresses the more widely disseminated of repeated climate scare reports. These are the common reactions to what are naturally occurring events. These have not intensified since the pre-industrial era.

Wither now?

> *'Are the worst enemies of society those who attack it or those who do not even give themselves the trouble of defending it?'* **Gustave Le Bon**[1]

The Government is bound by statute to procure the elimination of fossil fuel emissions by 2050.

The National Grid itself puts the cost of achieving this goal at £3,000,000,000,000 (£3 trillion)[2] . This cannot be regarded as a firm estimate. It would be idle to suppose that the cost will not far exceed that estimate. According to Bloomberg New Energy Finance's latest New Energy Outlook report (22nd July 2021) the costs of reaching net-zero globally by 2050 range between $92 trillion and $173 trillion over the next thirty years.

All this to be endured because of a belief and hypothesis that is contradicted by scientific evidence.

There has been an advent of hope, opportunity, security, education, health, long life, and sustenance such as would have been unthinkable just 200 years ago.[3] Are we to embark on the abandonment of

---

[1] Charles-Marie Gustave Le Bon was a leading French polymath whose areas of interest included anthropology, psychology, sociology, medicine, invention, and physics. He is best known for his 1895 work *The Crowd: A Study of the Popular Mind*, which is considered one of the seminal works of crowd psychology.

[2] National Grid "Future Energy Scenarios" July 2020.

[3] Professor Plimer Emeritus Professor of Earth Sciences University of Melbourne *"Heaven and Earth"* p 27 2009 Quartet Books.

the energy base that has sustained the miraculous modern age and the prosperity that has brought us, if we could but realise it, to the 'sunlit uplands'?

Yet all this is being done with no feasible alternative. Solar panels and wind are but chimeras.

Not until we resolve to adopt an energy policy that depends on hydrogen, electrolysis and nuclear power will we have energy that is virtually without limit, not dependent on foreign resources and free from emissions other than oxygen. Then only may our country be able to cease altogether to depend on fossil fuels. That should be the challenge of governments over this century. That should be the our legacy to future generations.

President Eisenhower warned[4] against the dangers of science becoming corrupted by a handful of powerful elites that did not have the advancement of scientific knowledge as their central goal.

We can but hope that the British public will come to realise what costs and burdens will be inflicted upon them by the project of Net Zero not only in the weight of financial exactions but in enforced dislocation and disruption of their customary lives and ways.

As yet it seems that the predictions of climate catastrophe have not raised public concern above those other conditions of life that it holds to be of greater moment and importance. A UN Sustainable Development Goals world survey in 2019 of 9.7m people ranked climate action as the least concern for humanity. A further survey in June 2021 ranked climate action 13th.[5]

Figure 1

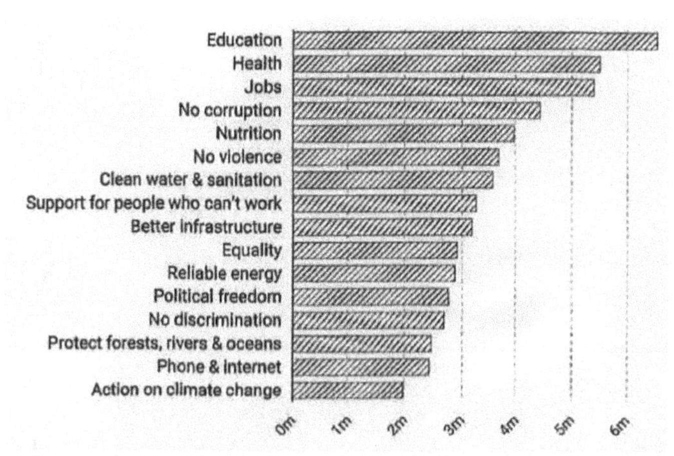

Who in public life will speak out for reason, impartial and rigorous weighing of evidence, and scrupulous scientific discipline? Surely we are not going to commit our country to expenditure matched only in times of war without at least a pause to allow deep and sincere examination of its justification?

Ultimately the dogma will be exposed by the passing of the years. Nature governs all. There is no 'climate change'. For 42 years now we have had daily unimpeachable observations and measurements from satellite and balloon radiosonde records testifying to this. As the shrieks of pending catastrophe intensify the evidence of Nature becomes inexorably incontrovertible.

---

[4] "Yet, in holding scientific research and discovery in respect, as we should, we must also be alert to the equal and opposite danger that public policy could itself become the captive of a scientifictechnological elite" Public Papers of the Presidents Dwight D Eisenhower 1035.

[5] Ipsos Mori poll 8th June 2021 "Global public ranks ending hunger and poverty and ensuring healthy lives as top priorities among U.N.Sustainable Development Goals.

PART 1

I CARBON DIOXIDE – ITS PROPERTIES AND VOLUME

Such has been the demonising of carbon dioxide by the IPCC and its disciples that the properties of this life-giving gas and its atmospheric scarcity have been altogether disregarded.

Properties

The use of the terms "net zero carbon", if that truly describes the intention of world governments, is misleading. There are 2 forms of natural carbon, graphite and diamond. However there are many carbon compounds. Two in particular are essential to life.

Calcium carbonate formed the multicellular sea creatures of the Cambrian explosion of life. It formed the South Downs. It is essential to formation of shells, skeletons, and corals.

Carbon dioxide is vital to all plants and trees. On the supply of CO_2 depends the entire biosphere. Plants and trees absorb CO_2 from the atmosphere. Photosynthesis transforms CO_2 into oxygen on which all animal life depends. Increases in CO_2 density promote increases in plant life. During the Early Eocene Climatic Optimum atmospheric content of the atmosphere was of the order of 1125 parts per million (ppm)[6]. This brought about a rapid expansion of vegetation. There were vast forests even in polar regions; new families of plants evolved to take advantage of the excess CO_2 which was created by an enhanced reaction between the atmosphere and rock resulting in chemical weathering.

The following diagrams show the impact of increases in CO_2 on agricultural yields in graph form[7] and global mapping. Atmospheric pollution arises from emissions of sulphur dioxide, carbon monoxide and other compounds. CO_2 is in no sense a pollutant[8].

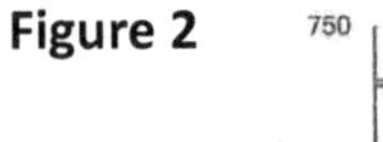

Figure 2

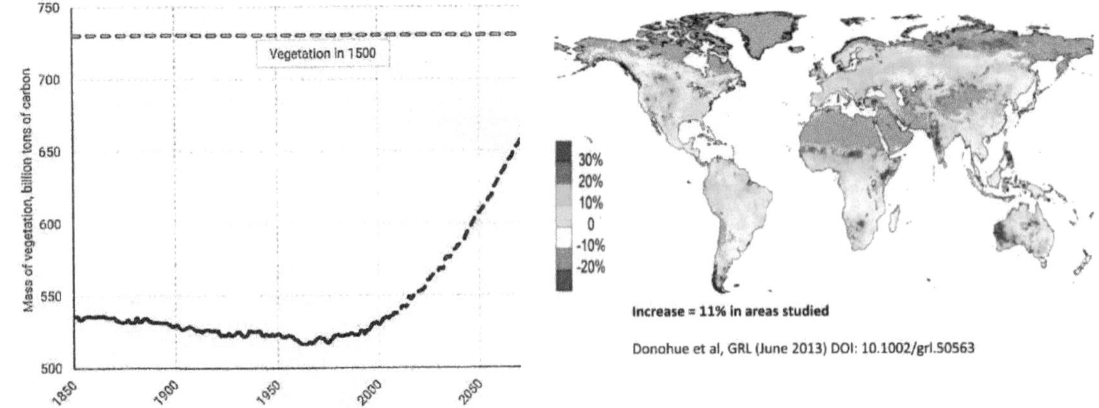

In periods of glaciation CO_2 levels fall drastically as a consequence of the acceleration of the absorbency of the oceans. Solubility of CO_2 in the oceans rises with falling temperature. In past epochs draw down of atmospheric CO_2 due to calcium carbonate shell and skeletal formation (Cambrian) and global forestation (Carboniferous) resulted in steep falls of density.

The following graph show that there was a point in the last glaciation – only 18,000 years ago – when CO_2 fell to about 180ppm. At below 150 ppm extinctions occur since plant life is extinguished as photosynthesis ceases. Increases in CO_2 leads to the flourishing of life on Earth.

---

[6] Professor Plimer "*Heaven and Earth Global Warming The Missing Science*" p 206. 2009 Quartet Books Ltd and authorities cited.

[7] Bjorn Lomberg "False Alarm" 2020 citing RCP8.5 emissions scenario: V.K Arora and Scinocca 2016. Vegetation n 1500 from Hurtt et al 2014.

[8] See for example US Environmental Protection Agency 9th December 2009 statement that CO_2 is harmful. A material part of human CO_2 emissions is from human breathing.

Volumes

Figure 3

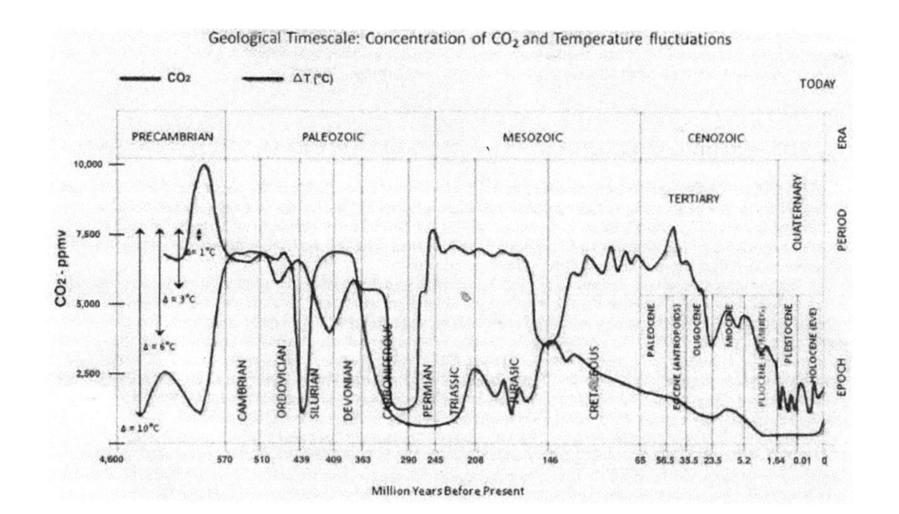

The Earth has had a fixed 'endowment' of carbon. It is today to be found in 'reservoirs'. The largest of these is the Earth's crust (1.9 bn gigatons[9]); then the oceans (40k gigatons); soils and living things (2.2k gigatons); underground fossil fuels (5-10k gigatons). The amount of carbon in the form of carbon dioxide in the atmosphere is just 2% of the content in the oceans. About half of the human induced CO_2 flow is taken up annually by the surface. Of the estimated 186bn tons of CO_2 entering our atmosphere each year **only 5%** is due to human activity. 57% is given off by oceans and 38% by exhalations of animals and humans.

CO_2 is reabsorbed by plants trees and the oceans. The lifetime of CO_2 in the atmosphere has been estimated at 5-10 years. Thus 18-10% of atmospheric CO_2 is exchanged each year[10]. However some studies show that the CO_2 is removed at a rate of only 2.3% each year over about 40 years. CO_2 density in the atmosphere has been increasing. For 10,000 years it has averaged between 260 and 280 ppm[11] until about 1850. It has risen since then to 410 ppm – an increase of 45% that is related to human activity. But global temperature has increased in that time by only 1.1%.

CO_2 represents 0.041% of the atmosphere today. For the last 5m years the CO_2 content of the atmosphere has been lower than the previous 550 million years except possibly in the Carboniferous-Permian era 300 million years ago. Atmospheric CO_2 has declined as a consequence of the 'scrubbing' of CO_2 by plants, forestation, oceanic absorption, and sequestering into sedimentary rocks. It is in geological terms lower than all previous epochs up to the Pliocene when hominids appeared.

CO_2 is a trace gas. It is a 'greenhouse' gas because it radiates reflected surface heat back to the Earth's surface but to a limited extent. CO_2 represents about 4% of these trace gases. Water vapour accounts for 95% - it is by far the most important and abundant greenhouse gas. Methane, nitrous oxide, CFCs, and a few others together account for about 1%. Section II explains how increases in CO_2 atmospheric density has a negligible effect on temperature once it reached approximately 250 ppm.[12]

SUMMARY

CO_2 is essential to life on Earth terrestrial and marine. Increases in CO_2 density are wholly beneficial since they result in the creation of surplus food, expansion of both vegetation and forestation. Any temperature rise due to increased atmospheric density is negligible.

---

[9] 1 gigaton = I billion tons.

[10] See the detailed exposition of Professor Plimer in "Heaven and Earth" pp 416 – 432 Quartet Book Ltd 2009 There are others who suggest longer periods – see Koonin "Unsettled" 2021 p 68.

[11] Ppm = parts per million by volume. Professor Plimer argues for a higher average Op Cit pp 424 et seq.

[12] *Relative Potency of Greenhouse Molecules*. Professor W A Wijngaarden Dept Physics York University Canada and Professor W. Happer.

II INCREASED CO_2 DOES NOT DRIVE MORE WARMING

Increases in density of atmospheric CO_2 do not also increase the infra red radiation it absorbs to more than insignificant extent. Doubling CO_2 will have negligible effect on temperature.

It is assumed by environmental activists that more CO_2 in the atmosphere means more global warming[13]. It is an assumption that informs the repeated statements of the IPCC and the catechism of the global warming 'consensus'. We are told that we are at a 'tipping point' which will be irreversible once passed. It would seem inevitable that if CO_2 density intensifies then its radiative effect would also intensify in proportion thus creating a runaway warming phenomenon.

Yet there is no correlation between a rise in temperature of the Earth and rise in CO_2 concentrations. This is explained in Section IV. Temperature has not increased in line with CO_2 either over geological ages or in the last 2,500 years. During severe glaciations in the mid Jurassic and early Cretaceous Period CO_2 density was eight times the present level.

How does this come about?

To explain it we need to understand that CO_2 is not climate sensitive over a certain density.

CO_2 Infrared wave lengths

The Earth maintains its temperature by balancing day time solar radiation (ultra violet short wave) and reflecting radiation from its surface (infra red long wave) both day and night.

Just under half of solar radiation is actually absorbed by the surface of the Earth as the rest is reflected or absorbed by clouds and the atmosphere before it reaches the surface. Earth surface heat is carried up into the atmosphere by evaporation and convection of warm moist air up to an altitude of about 7 miles (the lower troposphere). Above that height surface heat is transported upwards towards space by radiation.

The major atmospheric gases (oxygen and nitrogen) are transparent to incoming sunlight and also to outgoing thermal infrared radiation from the Earth's surface. However, water vapour, carbon dioxide, methane and other trace gases are opaque to certain wavelengths of outgoing thermal infrared energy. The Earth's surface re-emits[14] the net equivalent of 17% of incoming solar energy as thermal infra red radiation. But the amount that directly escapes to space is only about 12% percent of the incoming solar energy. The remaining fraction — a net 5%-6% percent of incoming solar energy — is transferred to the atmosphere when greenhouse gas molecules absorb thermal infrared energy radiated by the surface. This is in turn is radiated in all directions including back to the surface.

There are two important characteristics of CO_2.

In common with other greenhouse gasses including water vapour it only absorbs infra-red energy within certain wavelengths. In addition, adding more CO_2 to the atmosphere has a logarithmic effect and has a limit which is approached 'asymptotically'[15] limit in ever decreasing amounts.

Solar radiation is emitted at wavelengths of half a micron[16]. The thermal radiation from the surface of the Earth up to space is absorbed by greenhouse gasses according to the infra red wavelengths of the radiation. For CO_2 these are primarily in the bands 13 to 16 microns. The following graphs show the absorption wavelengths for CO_2. The left hand scale shows absorption potential expressed as a

---

[13] Goddard Institute for Space Sciences. For example *"Adding more CO_2 to the atmosphere is like putting another blanket on the bed"*.

[14] NASA Earth Observatory "Climate and Earth's Energy Budget" January 14 2009. 71% of solar energy (not reflected back by bright particles) gets through to the atmosphere and surface. The atmosphere radiates back to space (infra-red) the equivalent of 59% of incoming solar energy as to 23% absorbed by the atmosphere: 25% evaporation: 5% convection and 5 % - 6% thermal infra-red radiation absorbed by water vapour and other greenhouse gasses. The 12% balance of the 71% of total incoming is radiated direct to space.

[15] Pertaining to a limiting value for example of a dependnt variable when t he independent variable approaches zero or infinity.

[16] Micron = one millionth of a metre.

percentage up to total saturation. Three of the CO_2 bands are taken by H_2O. There are three available wavebands. But these are overlapped wholly or nearly all by water vapour bands. Water vapour is a far more efficient greenhouse gas than CO_2 with greater wavelength bands for absorbing radiation. At 1.9 – 2.1, 4 – 4.6 and at 13 – 16 there is a sharing of interception with CO_2 taking up what H_2O does not take up in the water vapour window.

Figure 4

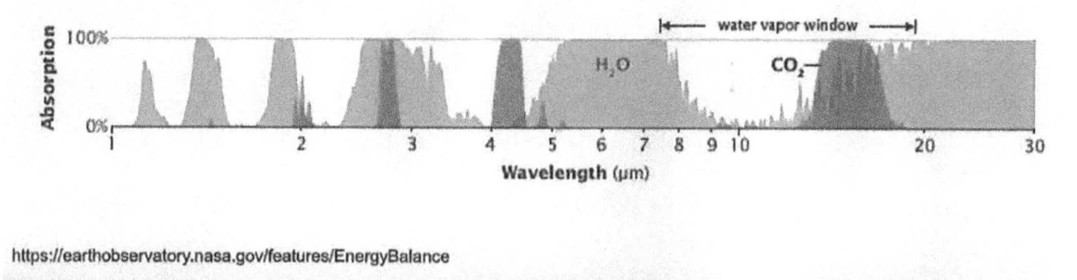

Absorption sensitivity

The absorption sensitivity of CO_2 is essentially logarithmic. Max Planck (1853 – 1947) was able to delineate the spectrum of radiation from warm bodies. He established what is known as the Planck Curve which appears as the blue curve in the following graph. The horizontal scale is the frequency of thermal radiation. The vertical scale is the thermal power going out to space. If there were no greenhouse gasses the radiation into space would be all that is comprised within the blue curve.

The actual infra-red radiation from Earth to space is described by the jagged black line in the following graph. The key element of this graph is the red line. This is what the Earth would radiate into space if the concentration of CO_2 were to double its current atmospheric volume.

Figure 5

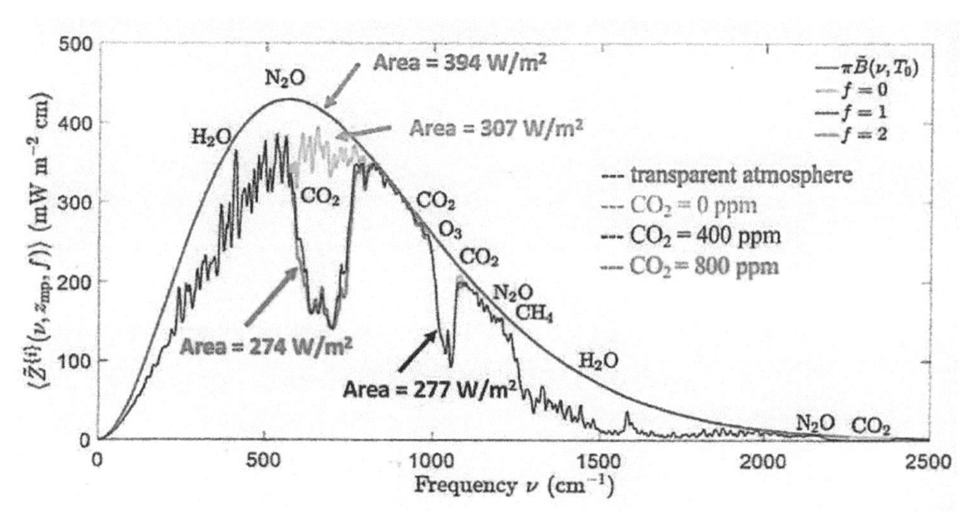

The green line shows the radiation from Earth to space in the hypothetical absence of all CO_2. The gap (bridged by the green line) is caused by CO_2 absorbing radiation from Earth that would otherwise go into space. If CO_2 density is doubled it does not double the gap. All that happens is a minuscule difference between the black curve (base density) and the red curve (double density).

Potency of greenhouse molecules

Earlier this year there was published a comprehensive analysis[17] of atmospheric 'forcing' - the effect of greenhouse gasses upon thermal radiation on a doubling of CO_2 density. It downloaded spectral lines and transition frequencies of over 1.5 million rovibrational[18] lines from the most recent

[17] *Relative Potency of Greenhouse Molecules*. Professor W A Wijngaarden Dept Physics York University Canada and Professor W. Happer Dept Physics Princeton University USA.

[18] Rotational–vibrational spectroscopy is concerned with infrared and Raman spectra of molecules in the gas phase. Transitions involving changes in both vibrational and rotational states are 'rovibrational' (or ro-vibrational) transitions. When these emit or absorb photons (electromagnetic radiation), the frequency is proportional to the difference in energy levels and can be detected by certain kinds of spectroscopy.

databases[19] in order to calculate the per-molecule forcing.[20] of the most important greenhouse gas molecules. Accurate calculations of radiative forcing were stated to be essential to any estimate of future climate change. The analysis examined the effect of changing greenhouse gas concentrations on thermal radiation in the case of a clear sky. Cloud cover would further reduce the forcing.

Its conclusion was that "*For current atmospheric concentrations the per-molecular forcings of the abundant greenhouse gasses H_2O and CO_2 are suppressed by four orders of magnitude*".

The absorption capacity of CO_2 is effectively logarithmic. The first 20 ppm of CO_2 operating as an atmospheric greenhouse gas has the most potent effect on temperature. After 250 ppm CO_2 has absorbed virtually all the infra-red radiation that it can absorb as shown in the following graph.

An analogy would be the use of sunglasses. The first pair blocks say 60% of sunlight. Two pairs block 24%. The next pair and it is 9.6% - the next it is 3.8% and so on. Thus the rise in CO_2 from 0.030% to 0.038% - just 0.008% - from 1950 to 2016 caused warming of only about 0.01%.[21].

Figure 6

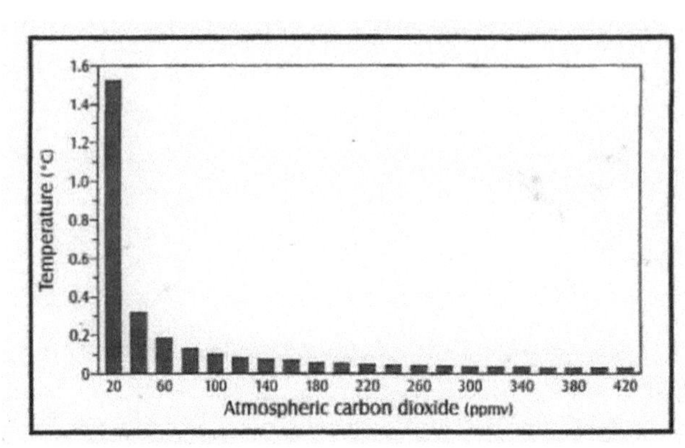

Doubling the concentrations of CO_2 to 800ppm accounts for only a further 0.8% reduction in cooling power[22]. Increasing concentrations of water vapour in the surface layer of the atmosphere caused by warming itself increases the proportion of absorption by water vapour of long wave radiation of the Earth's surface. This increase further reduces CO_2 absorption in the overlapping infra-red bands.

CO_2 accounts for no more than about 5% of the greenhouse effect. The rise from 1950 to 2021 of atmospheric CO_2 was just 0.011%. This cannot conceivably justify the predictions of climate modellers. It is said that increased water vapour is caused by minute warming from CO_2 with an indirect leverage of CO_2 greenhouse effect. This is just speculative. It is not derived from observational measurements and atmospheric water vapour has declined since 1948.[23]

SUMMARY

The addition of more CO_2 to the atmosphere will only result in negligible increases in its greenhouse effect and no further appreciable warming.

---

[19] HITRAN (an acronym for High Resolution Transmission) molecular spectroscopic database used to simulate and analyse the transmission and emission of light in gaseous media, with an emphasis on planetary atmospheres. VAMDC (Virtual Atomic and Molecular Data Centre) atomic and molecular (A&M) data compiled within a set of AM databases.

[20] A climate forcing is any influence on climate that originates from outside the climate system itself. The climate system includes the oceans, land surface, cryosphere, biosphere, and atmosphere. Examples of external forcings include surface reflectivity (albedo); human induced changes in greenhouse gases; atmospheric aerosols (volcanic sulphates, industrial output).

[21] Professor D.J Easterbrook "Evidence Based Climate Science" 2nd edn pp 166, 167. 2016 Elsevier Inc The attempt to show an amplification factor by reference to cloud cover and aerosols is merely speculative as explained above. Moreover cloud cover since 1948 has diminished.

[22] Koonin Op cit p 53 and Easterbrook pp 321, 322. See also *Relative Potency of Greenhouse Molecules*. Professor W A Wijngaarden Department of Physics York University Canada and Professor W. Happer Dept Physics Princeton University USA. Koonin Op Cit p 54.

[23] Professor D.J Easterbrook Op cit p167.

III THERE HAS BEEN NO ABNORMAL RISE IN TEMPERATURE

Measurement of the Earth's temperature provides the testimony on which judgment is made as to warming or cooling. So much is axiomatic.

How temperature is measured and where it is measured are critical to the validity of the data that is afforded by the measurement. The limitations of the measuring technique and the order of accuracy of that technique, as well as the way the data was analysed, are essential processes of science.

Satellite and balloon radiosonde observed measurements provide a gold standard of global temperature data.

Surface temperature records from weather stations are unreliable and are prone to manipulation.

Sources of data

Surface thermometer temperature records are maintained by two official entities. HadCrut is the Climate Research Unit working with the Met Office's Hadley Centre in Essex. The other is the Goddard Institute of Space Sciences[24] in New York.

Satellite and balloon sonde measurements are maintained by the University of Alabama at Hunstville under Professor John Christy and Professor Roy Spencer. Another satellite record is kept by Remote Sensing Systems though it is less comprehensive.

Since 1979 measurement of the atmosphere, including across the bands where the greenhouse effect occurs, has been made much more precise and reliable by introduction of satellite and balloon sondes. These cover the entire planet. They provide daily readings of different levels of the atmosphere. They are not concentrated in industrialised Western countries as are surface measurements. They cover the oceans equally as well as the land. The balloon radiosondes give a three dimensional profile of temperature. 'Renanalyses' are also provided by weather stations of meteorological and climate data to assimilate with atmospheric observational data.

Surface thermometer readings

Defects of surface records[25].

Surface temperature measurements derived from thermometers have serious limitations. They are greatly affected by local conditions including urban siting and the heat island effect[26], back reflections from structures, height above ground, changes in the size and materials of the required 'Stevenson Screen' – the box which houses thermometers. Approximately 50% of the US weather stations do not fulfil the requirements of the US government and introduce a warming bias

Loss of weather stations has impaired efficacy. In the mid 1970s there were 6,000 weather stations: there are now less than 1,500 with a disproportionate loss to rural areas. Most of those closed were cool, high altitude, high latitude, rural, and remote locations. The remaining stations are concentrated in populated industrialised Western countries and a much higher proportion are now in or are directly

---

[24] The Goddard Institute for Space Sciences was until 2013 run by Dr James Hansen. He is an impassioned believer in the dogma of global warming. With his colleague, the present director G A Schmid, he has been responsible for the GISS distortions of data described in this paper. His evidence to the US Senate Committee on Energy and Natural Resource in 1988 was that the 4 hottest years in 100 years recorded since the advent of recorded measurements had been in the 1980s rising to a peak in 1987. This was false. It put 'global warming' into the public arena. His statements are at the extreme end of environmentalist predictions of catastrophe including his warning to President Obamas that *"he had only 4 years to save the world"*. He called for CEO's of major fossil fuel companies to be put on trial for *"high crimes against humanity and nature"* for spreading disinformation about global warming. He described coal fired power plants as the *"factories of death"* and the *"trains carrying coal to power stations are death trains"*. *"Climate change is analagous to Lincoln and Slavery – Churchill and Nazism."*

[25] Plimer Op cit p 377 and authorities cited. See also Professor D Easterbrook Op Cit Chapter 2.

[26]Reflection from roads,pavements,domestic heating, factories, vehicles, machinery, shopping centres, power lines. T R Oke 1988. The Urban Energy Balance Progress in Physical Geography 12 471-508.

affected by built up areas or airports. Over 80% of the surface of the planet and 90% of the oceans have no coverage of instrumental temperature readings including vast areas of Russia, Africa, Canada, and Antarctica.

For these reasons the shortcomings of previous temperature measurements render them much more unreliable than sophisticated observation and measurement by satellite and balloon sondes. Surface temperature measurements should not be relied on if in conflict or inconsistent with these far superior observational readings.

Distortions of data by Goddard Institute of Space Science (GISS)[27]
and National Oceanic and Atmospheric Administration (NOAA)[28]

Moreover such measurements lend themselves to adjustments that convey a radically different trend to that dictated by the original data. This has been done by removing the sudden sharp temperature increase of the 1930s and the severe cold conditions from 1940s to the 1970s so as to produce a manufactured confection of approximate correlation with CO_2 density increase.

The matter of distortion of data to procure a steady lineal trend of warming since 1890 is considered in detail at Section VI in relation to the flawed processes of the IPCC. But it is important at this point to refer to these distortions when considering the actual IPCC records used to show temperature variations since then. That is because we now have 42 years of accurate and comprehensive daily readings from satellites and balloon sondes to compare with IPCC surface records for this period.

The two leading surface records have been the Goddard Institute of Space Science and the Met Office Hadley Centre working in conjunction with the Climate Research Unit – or HadCRUT. It has been repeatedly demonstrated[29] that both GISS and HadCRUT have deliberately altered data pertaining to temperature during pre-industrial times. This is done by adjusting temperatures downwards so as to make it appear that the rate of 20th century warming was greater than original data revealed.

By far the greatest concentration of surface temperature measuring stations is to be found in the US. By 1999 US surface temperature data had failed to show a steady warming trend and was showing a cooling trend from the 1940s. It also showed the universally accepted fact in the climate science community that the 1930s was by far the hottest period in recent times.

The average annual surface temperature data from the USHCNS[30] issued by NOAA were accordingly altered in 2007 from the NASA Goddard Institute of Space Sciences record in 1999 (left graph) which showed a cooling trend since the mid-1930s to show a steeper warming trend (right graph) .

Figure 7 **1999** **2007**

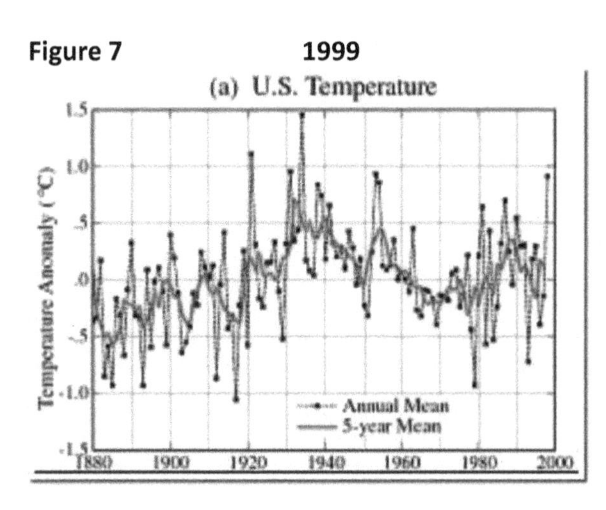

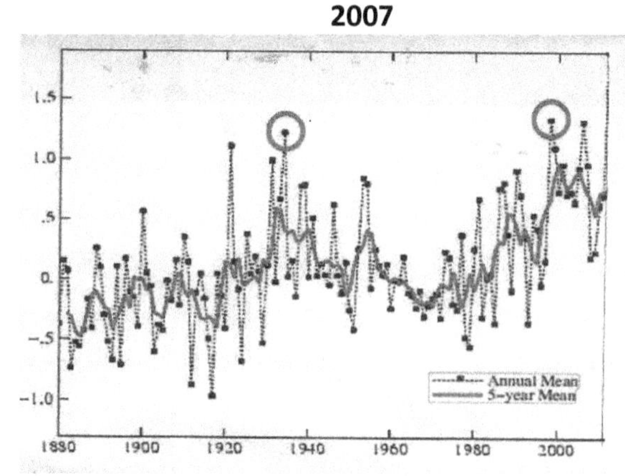

[27] Goddard Institute of Space Science.
[28] US National Oceanic and Atmospheric Administration.
[29] See e.g Dr William Soon, Heartland Institute. Dr Mototaka Nakamura "The Global Warming Hypothesis is an Unproven Hypothesis" 2019. Evidence-Based Climate Science 2nd edn 2016 D J. Easterbrook Professor Emeritus of Geology at Western Washington University.
[30] United States Historical Climatology Network Stations.

To achieve this data was simply changed. The surface temperature data now showed a much more pronounced warming trend with the 1930s written down drastically to create a linear average progression which did not exist with the original data and graph. This is shown summarised by Figure 8 which is a composite of Figure 7.

Figure 8

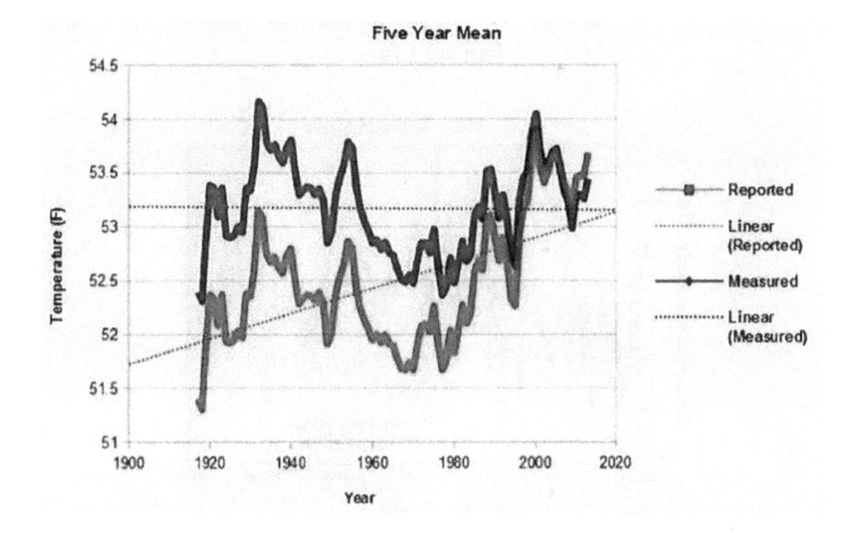

The blue line is the 5 year actual mean average annual temperature of all NOAA USHCN[31] stations. The red line is the published such average.

Misuse of data

The use of weather station data can also be easily distorted by taking the highest and lowest extremes and extrapolating a 'running record'. This enables a trend to be shown which depends only on showing one day which happens to be higher than a previous day.

Professor Steven Koonin[32] devotes a chapter in his most recent book to this distortion. His book is a detached and balanced work by a distinguished author. He addresses the flaws of the US Climate Science Special Report of 2017 (CSSR) which asserts that, based on an alleged running record, *"there have been marked changes in temperature extremes across the contiguous United States"*.

Figure 9

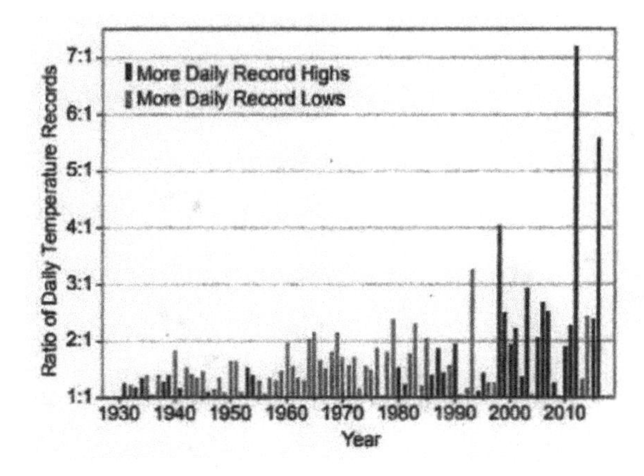

[31] Prof Easterbrook Op Cit p 50 and source cited.

[32] Chapter 4 *"Unsettled" What Climate Science Tells Us and What it Doesn't and Why It Matters"* BenBella 2021. Steven Koonin. He was Undersecretary for Science in the US Department of Energy under President Obama. Former professor of theoretical physics at Caltech. Professor at New York University. Author of 200 peer reviewed papers on physics and astrophysics. Trustee, Governor and holder of teaching posts at many established scientific institutions.

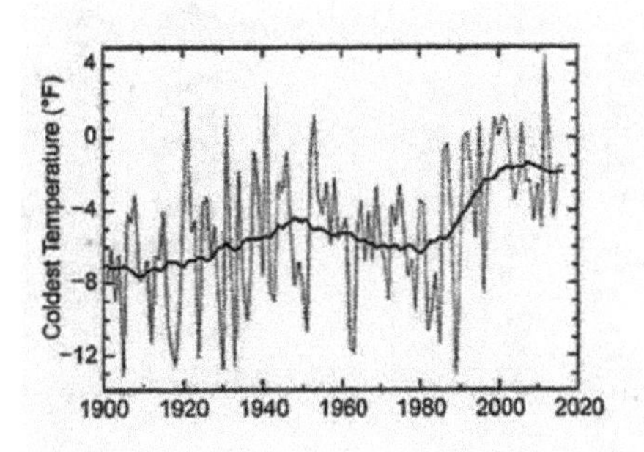

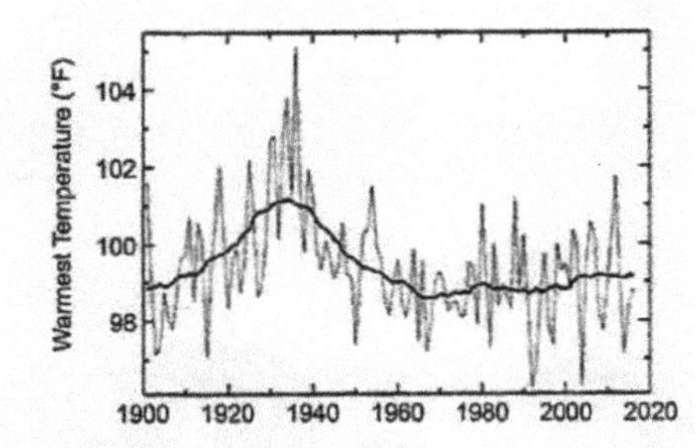

Professor Koonin shows how the CSSR use of running totals guarantees *"to show a long period of value around 1 at the start of the record, followed by dramatic variations at the end, creating the impression of large changes in recent decades even if they aren't present"*. This may be seen from Figure 9. The top graph is that set out prominently in the CSSR Executive Summary. It depicts the ratio of daily record high temperatures to daily record low temperatures across 49 states from 1930 to 2017. The lower graphs are those that appear in the actual text.

There were marked inconsistencies between this graph and the graphs in the actual text. Professor Koonin thought that the ratio of daily highs goes up because with rising temperature the number of daily colds diminishes: however the number of daily record warms remains constant. So the ratio changes severely with the denominator getting smaller against a constant numerator.

Professor Koonin and Professor John Christy [33]of the University of Alabama entirely reconstructed the US weather station data with code to analyse the data using absolute records not running records. Thus if the records changed over a period it would be on account of a change in temperature trend. Absolute record warms and colds were analysed from 725 US weather stations for the period 1895 to 2021.

Figure 10

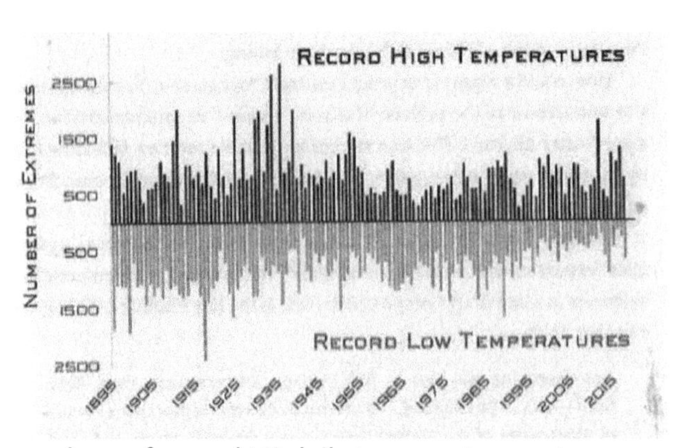

Figure 10 sets out the numbers of record US daily temperature extremes per 100,000 observations. It accords with known data as to the 1930s extremes of high temperature and also shows the cooling of 1950 – 1976 as well as the El Niño[34] spikes of 1998 and 2015. It is in line with the satellite data 1979 – 2021 as explained below.

Professor Koonin comments on the CSSR Executive Summary that *"there is no arguing that it is shockingly misleading"*. It is for all these reasons that Dr Nakamura (climatologist)[35] has concluded that global surface records *"no longer have any scientific value and are nothing but a propaganda tool"*.

---

[33] J.R. Christy Distinguished Professor of Atmospheric Science.
Alabama State Climatologist. Director of Earth System Science Centre Alabama University. Fellow of American Meteorological Society. NASA Medal for Exceptional Scientific Achievement.

[34] El Niño events occur roughly every two to seven years, as the warm cycle alternates irregularly with its sibling La Niña—a cooling pattern in the eastern Pacific—and with neutral conditions. El Niño typically peaks between November and January. Its effects take months to propagate around the world. El Niño is not caused by climate change. But it often produces some of the hottest years on record because of the vast amount of heat that rises from Pacific waters into the overlying atmosphere. Major El Niño events—such as 1972-73, 1982-83, 1997-98, and 2015-16—have provoked serious floods, droughts, forest fires, and coral bleaching events.

[35] Dr Mototaka Nakamura "The Global Warming Hypothesis is an Unproven Hypothesis" 2019.

Another comprehensive study of the best documented data sets that are not contaminated by bad siting of urbanization still concludes that such data sets *"are not a valid representation or reality"*[36].

Satellite and Balloon radiosonde measurements

The University of Alabama in Huntsville USA (UAH) publishes global temperature data each month derived from satellites orbiting daily and radio sonde data from balloons. This has been available since 1979. It is the most comprehensive body of observations of emissions from oxygen in the atmosphere. The intensity of these signals is in direct proportion to temperature.

The satellite instruments measure the temperature of the atmosphere from the surface up to an altitude of about eight kilometres above sea level. Very accurate records are obtained by measuring variation of microwaves from oxygen molecules. Balloon sondes report temperature at many levels. These sources provide reliable and consistent observed trends of atmospheric temperature. In addition planet-wide weather centres generate data on atmospheric conditions from satellites, balloons and other observations to create a continuously running model known as Reanalyses.

The UAH data show warming of the atmosphere since January 1979 of approximately 0.56°C (I°F). This is in the middle of the lowest sets of warming estimates put up by the IPCC which assume little CO_2 increase. But CO_2 density in the atmosphere has increased since 1978 by 80 ppm – that is by 24%. Moreover, for CO_2 to be a driver of surface warming the atmosphere where the greenhouse effect occurs must warm at a greater rate than the surface. But the opposite it the case.

Thus the most accurate and appropriate physical evidence does not at all support the dogma the CO_2 causes dangerous warming. It is instructive to review the data on temperature changes for the entire period of 42 years since 1979.

Figure 11

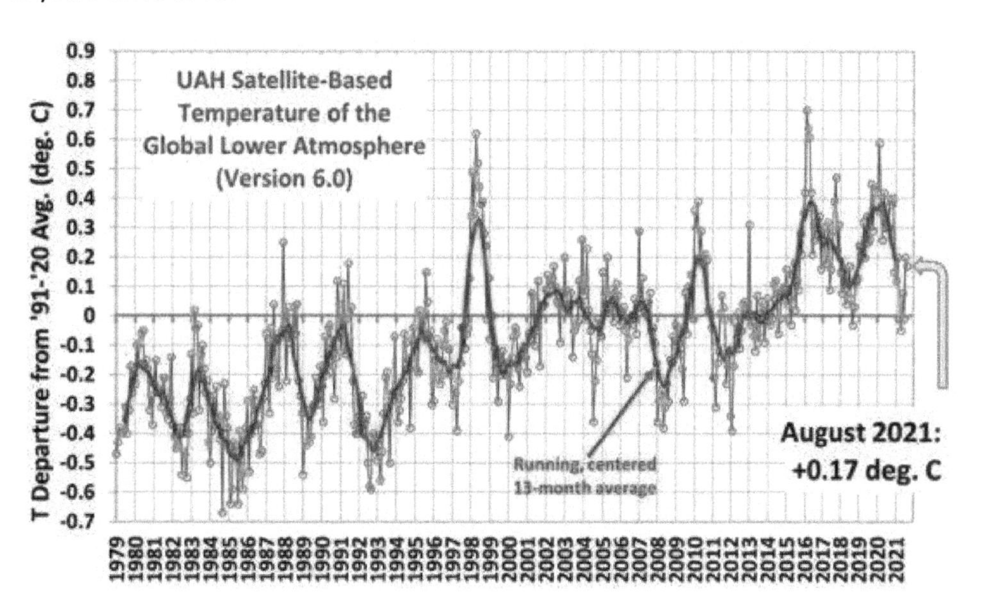

The data satellite/balloon record correlates with known naturally occurring phenomena. The dominant factors in the record are the El Niño spikes over these years. These included the El Niño spikes of 1982/3, 1987/8, 1991/2, 1997/8, 2015/6 and the effects of the more moderate El Niños of 1995 and 2019.

It also shows the effect of increased solar radiance due to high solar activity in 1979/80 in solar cycle 21 – the highest in the following 42 years. Average global temperature has not risen since 1998 and since then there has been a slight average fall.

---

[36] J P Wallace et al Idso 2017 "On the validity of NOAA, NASA and HadleyCRU average surface temperature data" et seq. Abridged research report – www.theresearch.files.wordpress.com.

How does this gold standard observational record of global temperature relate to the climate models used by the IPCC? To explain and illustrate this AUH[37] was able to set the actual trend of climate variation against simulated changes based on climate models[38].

The left hand graph show the average of 102 climate models of the IPCC derived from the CMIP5 models [39] used by the IPCC shown against the actual observed atmospheric temperature record from three sources of data – satellite, balloon, and reanalyses.

The right hand graph shows the Extended Reconstructed Sea Surface Temperature (ERSST) dataset[40] of ocean surface temperature and the CMIP6 climate model simulations which underwrite the IPCC 2021 'report'. The plot shows the monthly global (60N-60S) average ocean surface temperature variations from 1979 to 2016 for 68 model simulations from 13 different climate models. The 42 years of observations 1979 (bold black line) shows warming as occurring much more slowly than the average climate model says it should have. The rate of warming is 50% of what is being modelled. As seen from Figure 11 temperature has fallen since 2016 so widening the disparity.

Figure 12

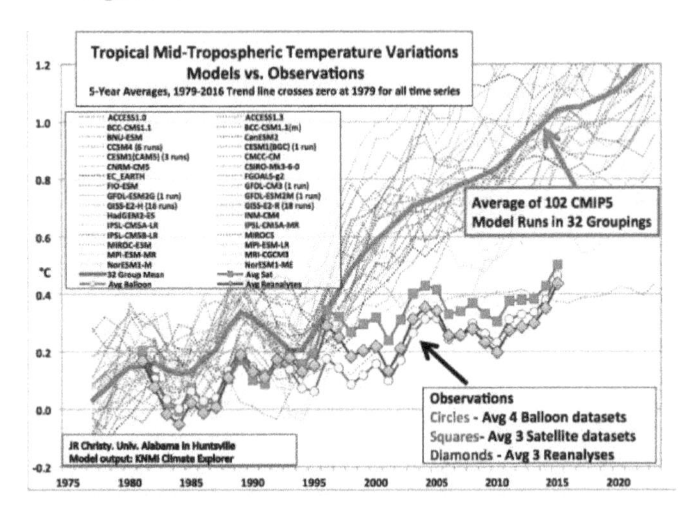

Figure 2: Five-year averaged values of annual mean (1979-2016) tropical bulk T_{MT} as depicted by the average of 102 IPCC CMIP5 climate models (red) in 32 institutional groups (dotted lines). The 1979-2016 linear trend of all time series intersects at zero in 1979. Observations are displayed with symbols: Green circles - average of 4 balloon datasets, blue squares - 3 satellite datasets and purple diamonds - 3 reanalyses. See text for observational datasets utilized. The last observational point at 2015 is the average of 2013-2016 only, while all other points are centered, 5-year averages.

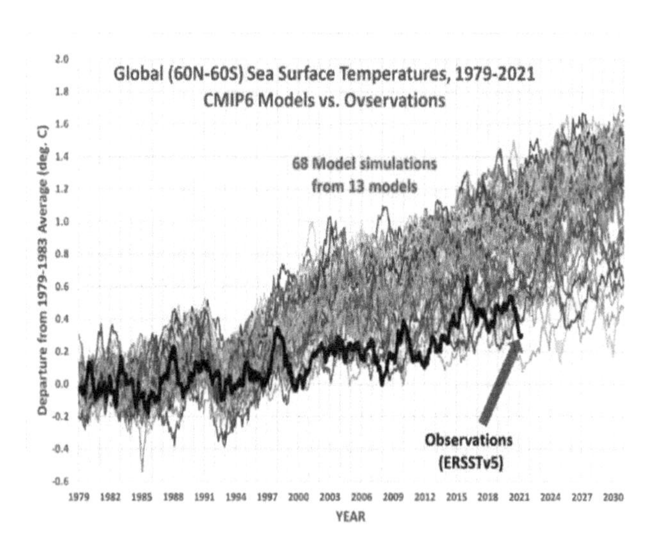

These graphs reveal most clearly two disturbing things:-

1. The models show a vast divergence from actual observational measurements. Evidence given at a hearing before the US House Science Committee[41] is that the model trend is *"Highly significantly different from the observations"*.
2. Furthermore it is obvious from these graphs that there is an astounding variation in the ranges of modelled temperatures. For example in the left graph there is so great a variation as to render averaging meaningless. Then again in the right hand graph consider the blue line at

[37] Alabama University of Huntsville. Official record of global satellite and balloon radiosonde temperature data

[38] Testimony of John R. Christy Professor Atmospheric Science University of Alabama to US House Committee on Science, Space and Technology 12th March 2017.

[39] Coupled *Model* Intercomparison Project Phase 5 The basis of IPCC modelling.

[40] A global monthly sea surface temperature dataset derived from the International Comprehensive Ocean–Atmosphere Dataset (ICOADS). Diagram Dr Roy Spencer University of Alabama *An Earth Day Reminder: "Global Warming" is Only ~50% of What Models Predict"*. The graph has been updated since data is still being released but the graph is not materially affected.

[41] Testimony of John R Christy Professor of Atmospheric Science University of Alabama to U.S House Committee on Science Space and Technology 29th March 2017.

the top right hand corner predicted for the next decade and its remoteness from the yellow line at the bottom right hand corner.

Conclusion

The IPCC model trends grossly exaggerate the actual observational trend. But when the models were run **excluding** the factor of added CO_2 they performed far better against the observations. This is due to the fact that increased CO_2 density in the atmosphere has negligible effect on temperature[42] (see Section II). The IPCC models algorithms are programmed with a CO_2 bias factor and it is this that creates the distortion. The models are false and of no value except to reveal their invalidity. A temperature increase of approx. 1.1^{0}C since pre-industrial time is substantially accountable by the fact that the planet has been emerging from the Little Ice Age the rise being attributable almost certainly to increase in solar activity (see Section IV Figures 21, 22 and 22.1).

SUMMARY

The IPCC has not proved that increase in global temperature since 1850 is caused by rise in CO_2 density and that it is not due to heightened solar activity on our emerging from the Little Ice Age. There has been no warming trend since 1998.

[42] Testimony of John R Christy as above. *Relative Potency of Greenhouse Molecules*. Professor W A Wijngaarden Dept Physics York University Canada and Professor W. Happer Dept Physics Princeton University USA...

Correlation[43] may simply be coincidence – it does not signify causality. However any theory of causation which does not give rise to correlation with the effect cannot be valid.

The lack of any causal correlation of rise in temperature with rise of CO_2 atmospheric density is examined both in geological time spans and also over the past 2,500 years.

423,000 years before present

Geology is concerned with the physical structure, substance and processes of the Earth.

A recently published re-appraisal[44] of ice-core indications re-affirms the crucial historical fact that CO_2 concentrations rise in response to increases in temperature. They do not cause temperature increases. For the past 4 ice ages, over 423,000 years, warming preceded increased density of atmospheric CO_2.

The Vostok Antarctica Research Station ice cores provide excellent samples of a fundamental feature of glaciation, namely the deglaciation cycles or 'interglacials' as they are often described. The intervals between interglacials are so long that the data record is not distracted by local or short span influences such as variations in the 11 – 1000 year solar cycles.

Vostok ice cores drilled down to 3,310 metres confirm the glaciation cycles as having time spans of between 87,000 to 123,000 years. Antarctic temperatures varied by 10ºC. Atmospheric density varied between 180 parts per million (ppm) and 300 ppm.

A further source of data is Dome C, 560 kilometres south of Vostok. It is the site of the European Project for Ice Coring in Antarctica (EPICA).

Vostok ice core analysis

Set out below is a graph depicting incidence of temperature changes and changes in CO_2 concentrations shown by the Vostok ice cores.

Figure 13

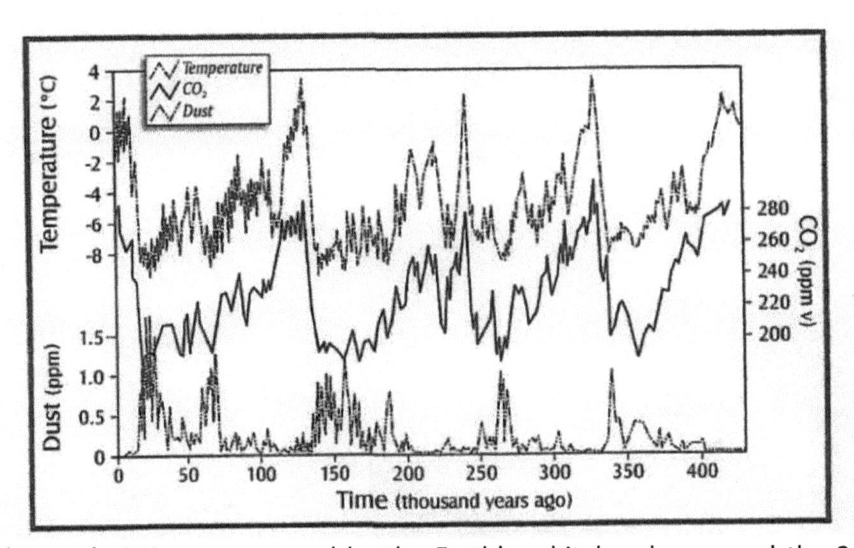

The glaciations and interglaciations are caused by the Earth's orbital cycles around the Sun – the eccentricity cycles of Milankovitch. Data is plotted in chronological order. On the graph the vertical bars denote the rapid and steep rises in temperature and CO_2 density at the outset of each interglaciation. The high peaks of temperature and declines are a consequence of the Earth's eccentric

[43] A mutual or reciprocal relationship between two or more phenomena.

[44] Pascal R Institut de Physique du Globe de Paris 26 May 2021 "The temperature – CO_2 climate connection: an epistemological reappraisal of ice-core messages".

orbital variations. The lowest of the three lines shows the levels of planetary dust in ppm. These correlate with descents into the cold glacial periods.

Temperature changes induce changes in atmospheric CO_2 and CH_4 (methane). The seas form 71% of the surface of the Earth. 58% of the Earth's CO_2 is in the oceans. The solubility of CO_2 in water varies greatly with changes in atmospheric temperature.

There is a striking proportionality between the extent of temperature increase and higher concentrations of CO_2. At the beginning of each cycle the range of rise and fall of temperature is over periods of approximately 7,000 to 16,000 years. For CO_2 the range is from 14,000 to 23,000 years. There is a lagging period of about 7,000 years. Similar lagging behind and spread is seen with the less severe peaks and falls but not nearly so wide – l,000 years is probably the average. Temperature rises cause an overall decrease in solubility in the ocean so releasing CO_2. This takes place over these lagging periods.

It is also evident that in times when temperature is falling CO_2 concentration does not immediately fall but declines relatively gradually as the solubility of the oceans increases.

Yet a further aspect of the ice core data is confirmation of desertification being a consequence of changed conditions at times of cooling and not of warming. Decrease of CO_2 directly diminishes photosynthesis and falls in temperature diminish evaporation. Both such factors govern desertification[45]. This is reflected in the dust record set out in Figure 13.

Attempts have been made to show that such rises in temperature were indeed caused by CO_2 increases. Such a hypothesis is inconsistent with the carbon cycle. It is accepted now that deglaciation occurs after an era of glaciation by reason of the eccentricity of the Earth's orbit round the Sun[46]. The most abundant source of CO_2 by far in the carbon cycle is the oceans. No other source of CO_2 could possibly account for the increases in CO_2 density revealed by the ice core data. CO_2 is absorbed by the oceans in cold periods which is evident from the CO_2 levels during glaciation. It is released with warming of the atmosphere.

The long term global average of sea surface temperature is 15^{0}C. At this level seawater can dissolve its own volume of CO_2. However at 10^{0}C it absorbs 19% more and at 20^{0}C it absorbs 12% less. What happens during 'degassing' is that cold polar deep ocean water moves with dense bottom currents to tropical latitudes where it upwells to the surface releasing CO_2 as it warms. 70% of degassing takes place in this way.

There is a definite correlation between CO_2 and temperature disclosed by the Vostok ice cores which is inconsistent with the consensus dogma of global warming. It is consistent only with the existence of quite another cause of CO_2 increase – the orbital cycle of the Earth and the consequent rise and fall of temperature releasing CO_2 and absorbing it.

EPICA Dome C ice cores

The ice core data record from the EPICA Dome C of Antarctica reveals temperature variations and CO_2 densities from 20,000 BP. Temperature rise started thousands of years before CO_2 concentrations in the atmosphere increased.

---

[45] See Plimer Op Cit pp204 – 207 and authorities cited.

[46] One of the Milankovitch cycles.

Figure 14

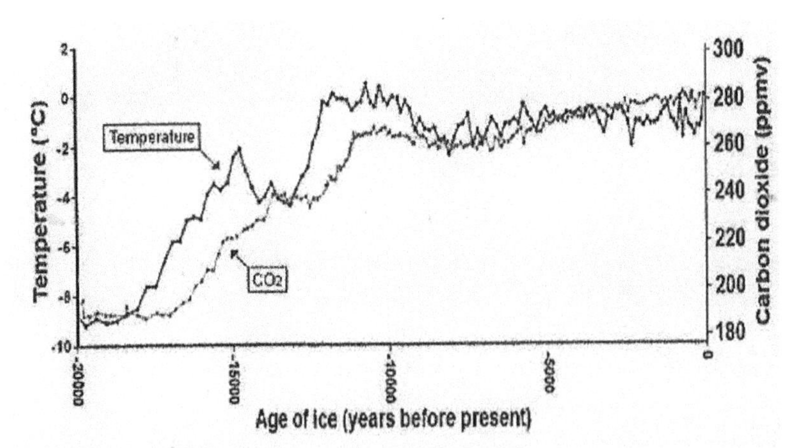

There is a correlation but it is one which is utterly inconsistent with the hypothesis of CO_2 inducing a rise in global temperature. For millions of years in severe glaciations CO_2 has been over 8 times the present density and even higher in the Pre -Cambrian. The Dome C ice core data confirm that rises in temperature ceased before CO_2 density rose. For all of the graph's 20,000 years CO_2 has not driven global warming. The inverse is the case.

Timescales of 150 years are laughably inadequate to base any reliable conclusions as to the phenomena that govern climate variations. It is only by reference to the causes of de-glaciation and glaciation, orbital cycles, solar radiation, and atmospheric conditions that any useful scientific conclusions can be proposed.

All five studies of the Antarctic ice cores[47] showed CO_2 lagging behind warming over a period from 420,000 years ago. There is no causal correlation with the pronounced variations of temperature in that time.

Last 2,500 years

Temperature fluctuations have been a characteristic of climate during the Holocene era in which we find ourselves. The Greenland oxygen isotope records provide a reliable record. The following graph[48] reveals that there were at least 22 temperature fluctuations between 1480 and 1950 when CO_2 levels started to rise steeply. None of these could have been caused by increased CO_2 as all preceded the rise of CO_2 density levels.

Figure 15

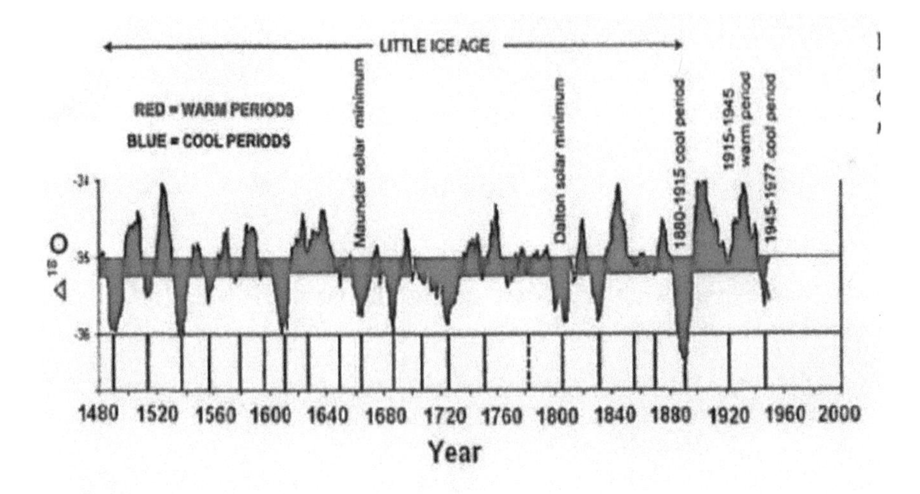

The principal global temperature variations for the past 2,500 years have been the Roman Warming Period, the Medieval Warming Period and the Little Ice Age. During each of these periods extending

---

47 See Prof Easterbrook Op cit pp 170 171.

48 See Prof Easterbrook Op cit p 211citing Stuiver. M et al 2000 GISP2 oxygen isotope ratios. Quaternary Research 54 (3).

over 2,500 years CO_2 atmospheric concentrations have remained within a range of 260 – 290 ppm. These periods saw pronounced variations of temperature.

There is one simple correlation between temperature and CO_2 atmospheric density. Between 1550 and 1800 there was a significant relative fall in CO_2 concentrations during the Little Ice Age when solar activity was negligible and a very cold temperature prevailed. The decline in atmospheric CO_2 occurred with the steep temperature drop over the previous 250 years that caused acceleration of absorption of CO_2 by the oceans.

There is no causal correlation whatsoever with any rise in temperature.

These periods of warming deserve examination as the IPCC has sought to eliminate them from their graphs. This has been done to enable it to assert a causal correlation of temperature rise and CO_2 intensification. It was done in the Summary for Policy Makers of the 2001 'report'. It has now been done in the 2021 'report' and so serious is this deliberate distortion that it is examined in detail in Section VII below.

Roman Warming Period

In 2020 there was published a new surface sea temperature reconstruction from the Sicily Channel[49] based on magnesium/calcium ratios measured on the planktonic foraminifer *Globigerinoides ruber*. This creature deposits its calcium carbonate close to the oxygen isotopic equilibrium. This equilibrium is temperature sensitive making it a reliable means of establishing past changes in the temperature.

The record of data (left hand graph) is set out in comparison with previous surface sea temperature reconstructions from the Mediterranean. The study was framed in the context of previously published records from the Alboran Sea, Minorca Basin, and Aegean Sea. This comparison is complemented with a reconstructed NAO[50] record and also with a record of north Hemisphere temperature variations (right hand graph). The study confirms the persistent regional occurrence of a distinct warm phase during the Roman Period. These records consistently show the Roman Warming with the Medieval Warming as the warmest periods of the last 2,000 years – at least 2°C warmer than over current average values.

Figure 16

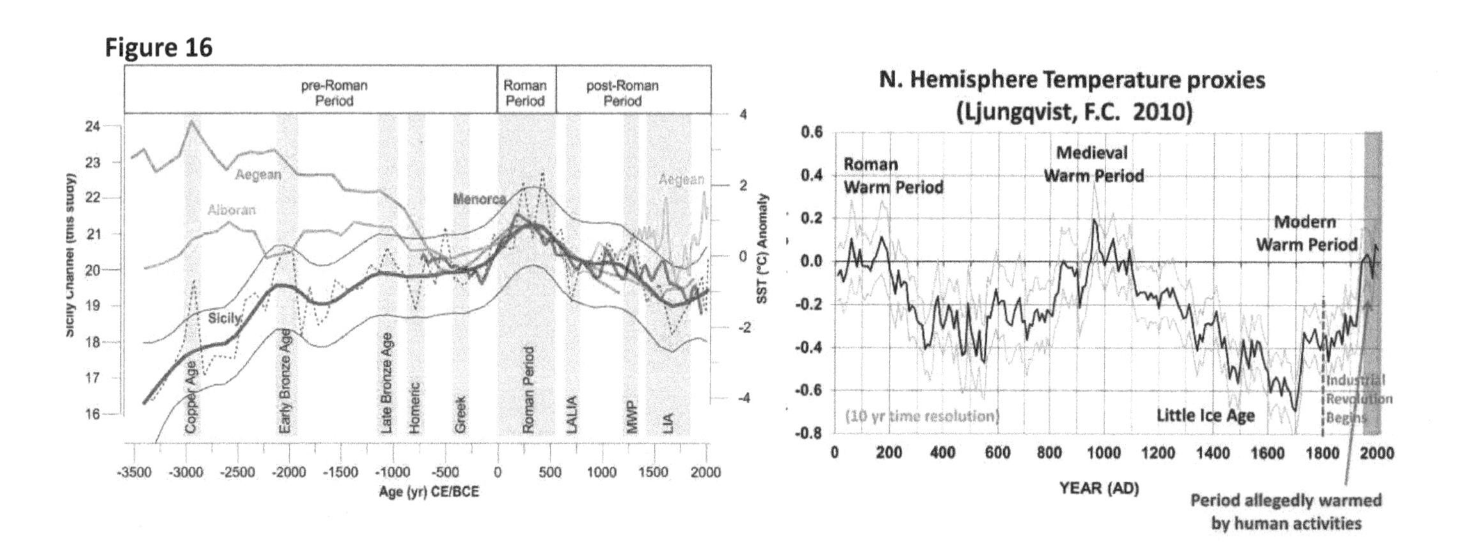

[49] Margaritelli, G., Cacho, I., Català, A. *et al.* Persistent warm Mediterranean surface waters during the Roman period. *Sci Rep* **10,** 10431 (2020). https://doi.org/10.1038/s41598-020-67281-2.
[50] US National Oceanic and Atmosphere Administration.

Medieval Warming Period

There are many regional proxies[51] from around the world that establish the existence of a warm period almost universally warmer than today called the Medieval Warm Period or Medieval Climate Optimum from 900AD-1350AD. The North Hemisphere graph Figure 16 above shows this period most clearly and depicts how temperatures exceeded those of today.

There is strong evidence from studies of the sea bed in the Sargasso Sea by the Woods Hole Ocenographic Institute. The Sargasso sea bed has high sedimentation which produces a surface dwelling plankton *Globigerinoides ruber* which, as described above, deposits its calcium carbonate close to the oxygen isotopic equilibrium. This equilibrium is temperature sensitive making it a reliable means of establishing past changes in the temperature of Sargasso Sea surface waters.

Figure 17

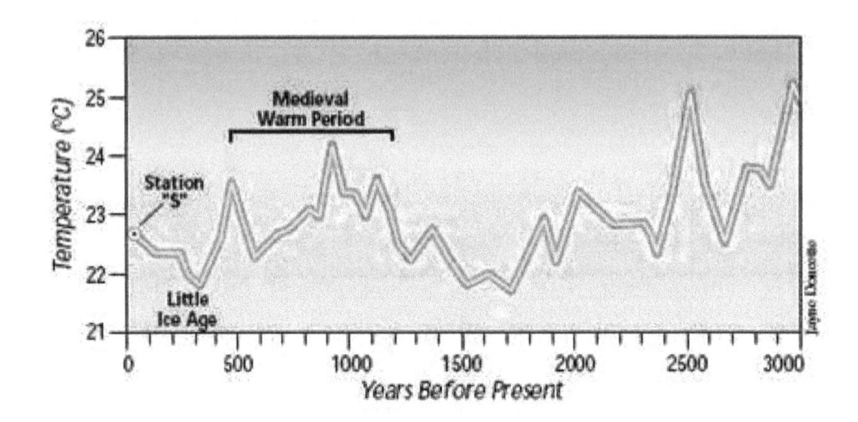

Note that not only can the Medieval Warm Period be seen (1,200 – 600 years BP) and the depths of the Little Ice Age (400 – 170BP) but also the Dark Ages cold period (1900 -1200 years BP) and the Roman Warm period (2,600 - 1900 years BP). Moreover the Medieval Optimum was a global phenomenon as appears from the following graph[52] of Northern Hemisphere temperature data.

Figure 18

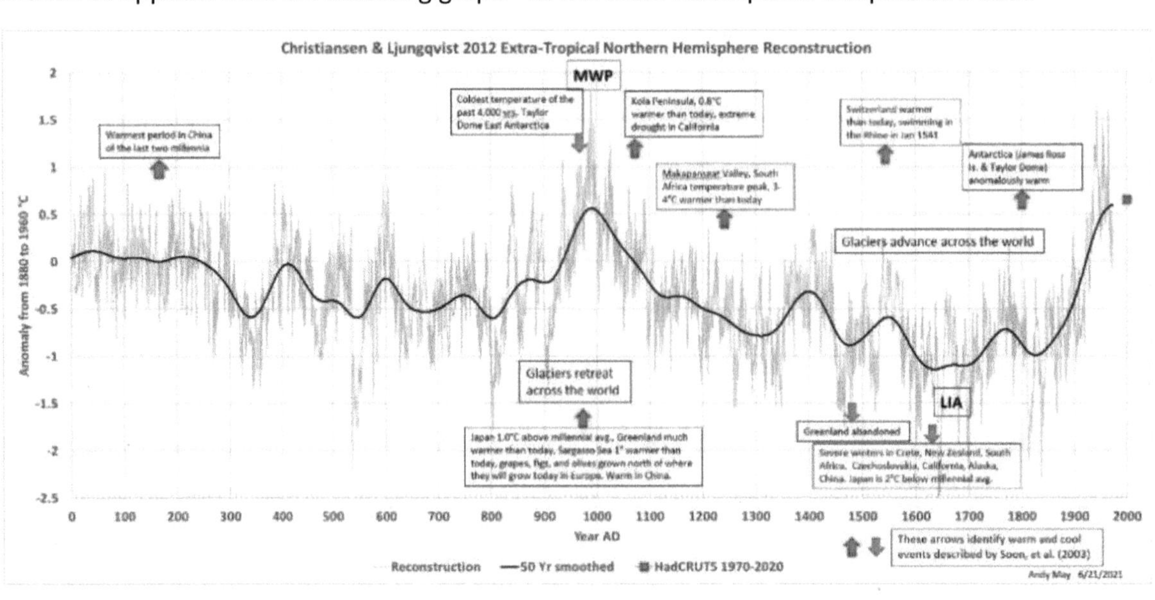

---

[51] Proxy data comprise sources of temperature indications other than direct measurements. They are physical characteristics of the environment preserved to varying degrees. Proxies are natural indicators of climate variability and include data from sources such as tree rings, ice cores, fossil pollen, plankton, ocean sediments, and corals. These allow reconstructions to be made of estimated temperature levels.

[52] Christiansen, B. and Ljungqvist, F. C.: The tropical Northern Hemisphere temperature in the last two millennia: reconstructions of low-frequency variability, Past, 8, 765–786, https://doi.org/10.5194/cp-8-765-2012, 2012.

Last 150 years

CO_2 concentrations in the atmosphere have increased since 1850 to the present day from 280ppm to 417ppm. A gradual rise until 1950 was followed by steep acceleration in a linear curve.

The surface temperature records from weather stations are so unreliable, due to the distortions described in Sections III and VII, that only graphs from periods prior to the beginning of the global warming scare (1983) are likely to be in their original form. The graph on the left shows CO_2 density rising since 1880. It is that published by NOAA[53]. The graph on the right is that published by James Hansen[54] in 1981 in the year that he became director of GISS.[55] This was at a time when global cooling had been predicted for over three decades.

Figure 19

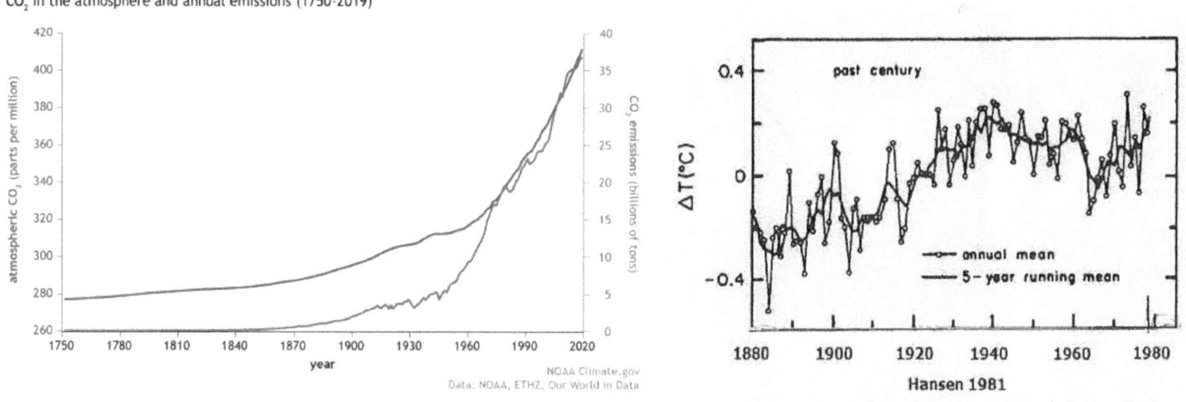

For the 100 years of the graphs to 1980 it is evident there is no causal correlation between the rise of CO and of temperature. As has been shown in Figure 11, since 1980 there has been a rise in temperature of approximately 0.56°C but with no more average rise over the last 23 years.

The lag of CO_2 behind rise of temperature above over vast spans of geological time has been noted. However a further most important factor is the fact that CO_2 lags behind temperature increase over even very short time spans

Figure 20[56]

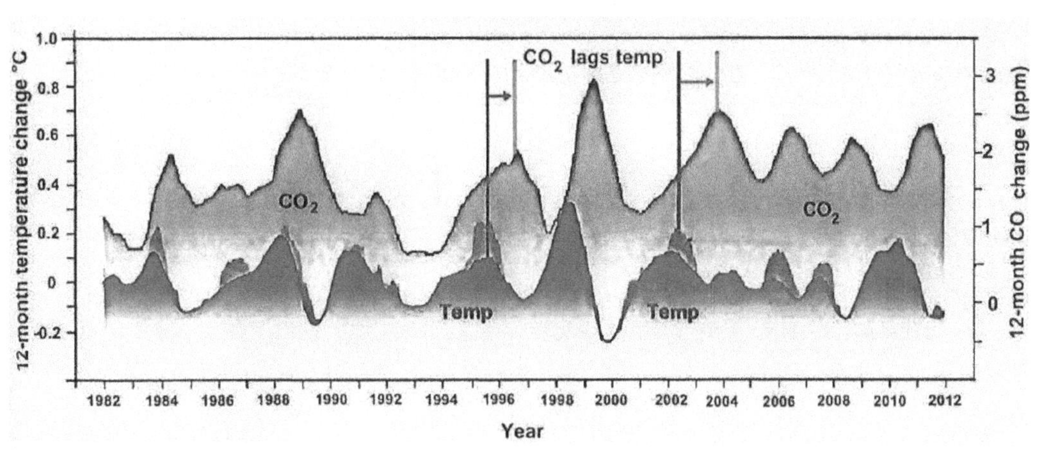

That being the case it cannot be that the rise in temperature is caused by the later CO_2 rise.

---

53 US National Ocean and Atmosphere Administration.

54 James Hansen is regarded as the progenitor of the global warming dogma.

55 Goddard Institute of Space Studies. One of the official custodians of surface temperature measurement.

56 Humlum. O et al. University of Oslo "Identifying natural contributions to late Holocene climate change" 2011, 2013 Global and Planetary Change 70. 145-156 cited by Professor Easterbrook Op cit p 173.

Correlation with Solar output

Whilst there is no correlation of temperature increase with a rise in CO_2 concentrations there is a definite and persistent correlation with solar activity[57].

Cycles of solar radiance are of various durations. Ice sheet cores and numerous other proxies reveal that they are of 11, 22, 87, 21 and 1500 years. The IPCC climate models do not take account of the critical importance of solar cyclical radiation and none predicted the early 21st century cooling.

The causes of temperature rise depend on what scale of time is being considered. A scale of millions of years reveals that the eccentricity of the Earth's orbit round the Sun governs the cycles of glaciation (ice ages) of approximately 90/120,000 years and the short succeeding interglacial periods of approximately 10/14,000 years.

On a shorter scale the variations in radiative intensity of the Sun correlate with Earth's temperature changes to a significant degree. This is the primary factor that accounts for the temperature rises of the Roman Warming Period (250BC – 450AD), the Medieval Warming Period (950AD – 1300AD) and the modern warming. It also accounts for the Little Ice Age (1310AD to 1850AD).

Figure 21

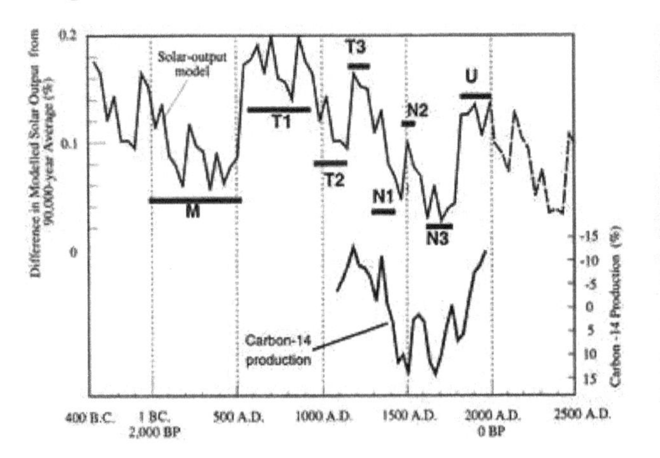

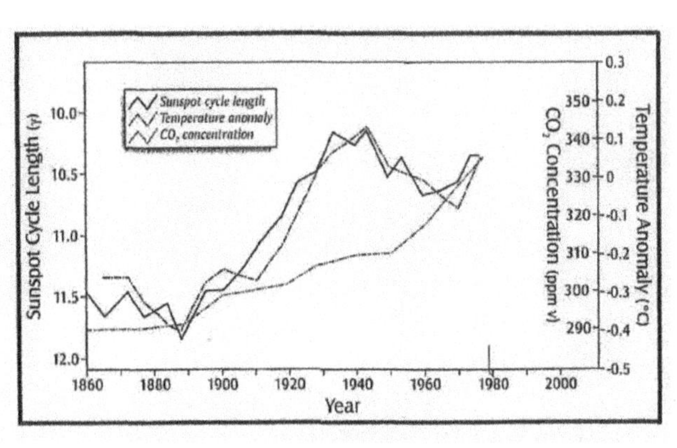

In the left hand graph there is correlation with the Roman Warming Period which opens the record. The next warm period is the Medieval Optimum (points T1 – 3). That is succeeded by the Little Ice Age lasting (points N1 – N3) when there was no material solar activity. There follows the temperature rise from 1850 (point U) to the end of the last century.

Currently the Earth is enjoying the latest warm period (point U) which has been with us for almost two centuries with a modest rise in temperature of 1.1°C.

The second graph (right) shows the 20th century solar activity to 1979 related to temperature rise and the CO_2 gradual ascent. Both graphs show the 1930s spike and the 1945 – 1976 fall - with great precision in the 20th century graph where the temperature and solar activity closely correlate.

Dr Soon[58] has demonstrated how the Arctic temperatures (Polyakov) correlate extremely well with the total solar irradiance. He also shows that there is no correlation with CO_2 densities.

---

[57] Solar natural phenomena occurring within the magnetically heated outer atmospheres in the Sun. These phenomena take many forms, including solar wind, radio wave flux, energy bursts such as solar flares, coronal mass ejection or solar eruptions, coronal heating and sunspots.

[58] Dr W Soon astrophysicist at the Solar, Stellar and Planetary Sciences Division of the Harvard-Smithsonian Center for Astrophysics. Founder of Center for Environmental Research and Earth Sciences (CERES-science.com).

Figure 22

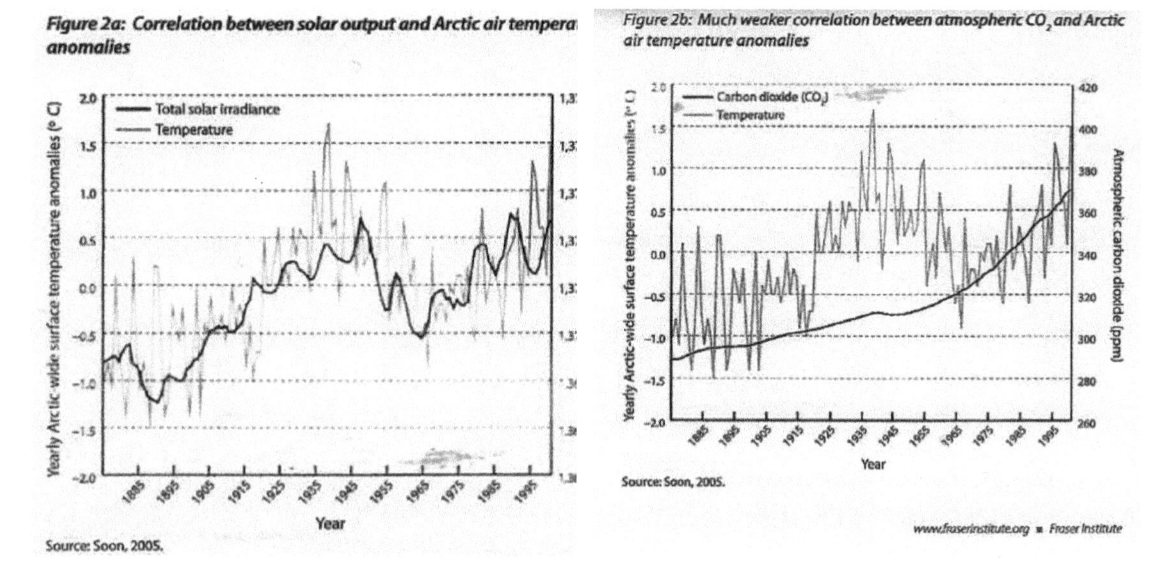

All of these close correlations are so striking as to be compelling.

They provide strongly persuasive evidence that the driver of climate change is solar activity. They illuminate the rise and fall of temperature over 2500 years.

Solar activity has declined over the past two cycles[59]. The last cycle (Cycle 24) from 2008 to 2014 was the weakest for 100 years. This accounts for the static and slight downward trend of temperature over these years. The current cycle seems to be at a similar lower level of intensity. If that is a trend then we can expect noticeable cooling in the future.

Figure 22.1

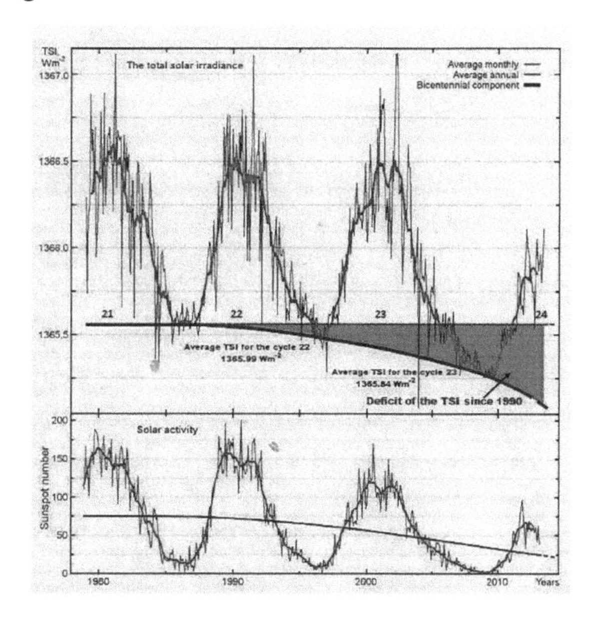

SUMMARY

There must be a necessary correlation between the cause of a phenomenon and its effect. There is close correlation of temperature with solar activity. The absence of any causal correlation of rise in temperature and CO_2 density invalidates the hypothesis that CO_2 is the driver of global warming.

[59] Professor Easterbrook Op Cit p311, Frohlich 2013.

PART 2

V ENVIRONMENTALISM & DOGMA

"In questions of science the authority of a thousand is not worth the humble reasoning of a single individual" ***Gallileo Galilei***

"Let him not be taught science, let him discover it. If ever you substitute authority for reason he will cease to reason; he will be a mere plaything of other people's thoughts".
Jean Jaques Rousseau *On Education (1762)*

All of us care for our planet. To deny this is to ignore the wonder we all feel when we regard the beauty of the Earth revealed in the Earthrise image of the Apollo 8 mission.

It is a simple disposition not to inflict harm as far as it reasonably lies in our power. We would like to make our planet a better place. We have a love of Nature and her beauty. After the toils of earlier times we seem to have come to the sunlit uplands.

We also cherish the freedoms that have taken generations of struggle to achieve. We wish to lead a peaceful life of co-existence. Whatever the struggles and sorrows we may endure we cannot but be grateful that we live in what is surely the best age there has been for humanity.

How is it that all this seems threatened by the catastrophe and doom predicted by those who, by preying on our natural concern, have induced an underlying state of guilt and fear and caused us to embark on truly fundamental changes to our settled form of life?

To gain an understanding of how this has come about we must see how a simple and deep instinct of care for our planet has been traduced and exploited by the dogma of modern environmentalism that has been preached to us ever more stridently for over 30 years.

It is a means of procuring by intimidation the overturning of the basis of our industrial economy.

The nature of environmentalism

Environmentalism is in the nature of a religion. It depends on belief. It does not depend or vary according to verifiable and compelling evidence. It is maintained despite such evidence. This belief derives authority by claims that it is held by all right thinking people. That is the weight of 'consensus'. It is the force of an assumed common belief.

Because it is a belief its adherents must defend it. Contrary opinion is not admitted. Those who doubt or challenge the belief are denied dissemination of their views. They are vilified. No dialogue with them is permitted or held to be possible. It is a dogma.

Dogma consists of a principle or theory or set of principles or theories laid down by an authority as incontrovertibly true. It is so far beyond dispute or argument that the views of those who advance contrary principles must be suppressed and their advocates scorned and excluded.

These are the principal elements of climate environmentalism – Belief, Consensus, and Dogma.

Belief v propositions of science

As one of the founders of Greenpeace stated,

"It doesn't matter what is true. It only matters what people believe to be true".[60]

This belief was essentially that a global revolution was necessary to end disparities between the Northern and Southern hemispheres, arrest the progress of modern economies, dismantle industrialisation and re-distribute wealth. This could not be achieved by democratic process through nation-states. A way had to be found to transcend historical systems of governance using a supranational body of prestige and authority to predict global catastrophe which could only be averted by a global overturning of the energy base of modern economies.

Origins of environmentalism

Environmentalism as a socialist form of overthrow of the global status quo was launched by a small group based on Stanford University USA centred on Stephen Schneider Professor of Environmental Biology and Global Change in the early 1970s. It gave impetus to the Club of Rome which first enunciated the environmentalist dogma. It was Schneider who declared:-

> *"To capture the public imagination we have to offer up some scary scenarios, make simple dramatic statements and little mention of any doubts one might have. Each of us has to strike the right balance between being effective and being honest[61]"*

The means by which the believers would procure a new order was described in the blueprint for the 21st century in published in 1990. It has a sub-title "A Report by the Council of the Club of Rome". It deals with the basis of environmentalism and explains that to achieve the sought for global change new enemies had to be found. It notes that, historically, social or political unity has commonly been motivated by identifying enemies in common,

> "*The need for enemies seems to be a common historical factor... In searching for a new enemy to unite us, we came up with the idea that pollution, the threat of global warming, water shortages, famine and the like would fit the bill....All these dangers are caused by human intervention and it is only through changed attitudes and behaviour can they be overcome. The real enemy then is humanity itself."*

The need for a 'new enemy' was essential to the promise of salvation only by fundamental re-casting of economic activity and wealth distribution. The creation of the vision of a global apocalyptic catastrophe which could be averted only by global means.

The vital components were fear and guilt. The 'enemy' in the post war era had been impending nuclear war and the folly of mankind. This abated with the collapse of the Soviet Union but it had cast humanity as its own enemy. It was thus in keeping with the times for environmentalism to demonise humanity in order to create guilt to compound fear. Such is the message contained in the 'Second Report to the Club of Rome' in which it is declared,

> *"The Earth has cancer and the cancer is Man."*.[62]

Thus it was that dogma of global warming caused by humanity was conceived.

The Dogma of 'Climate Change' became the force that it is today by casting over its political ends a cloak of plausible scientific claims, repeatedly asserted. That this was achieved is largely by the life's work of one Maurice Strong[63]. Strong derived inspiration for his ideas from the Club of Rome which

---

[60] Paul Watson cited in "The not so peaceful world of Greenpeace" Forbes Nov 11 1991 and in "Unsettled" Steven E Koomin BenBella 2021.

[61] Dr Stephen Schneider "Discover" October 1989.

[62] Mesarovic, Mihajlo; Pestel, Eduard (1975). *Mankind at the Turning Point*. Hutchinson. ISBN 0-09-123471-9.

[63] 1929 – 2015 President of Power Corporation of Canada until 1966: early 1970s Secretary General of the United Nations Conference on the Human Environment: then first executive director of the United Nations Environment Programme. Chief Executive Officer of Petro-Canada from 1976 to 1978 and chaired Ontario Hydro, one of North America's largest power utilities. He served as a commissioner of the World Commission on Environment and Development.

had published 'Limits to Growth' in 1972. It predicted collapse of the modern economies and the environment due to population growth and unavailability of natural resources.

It is now accepted that its arguments in support of these predictions cannot be sustained and none of them have come about. However, among the Club of Rome's doctrines was that destructive development had been brought about by the use of fossil fuels which must now be eliminated.

Strong, from his earliest years, was obsessed with the notion that the goal of mankind should be a world government. Born in 1929 he explained in later life that he was radicalised by his experience of growing up in the Depression. He described himself as a 'Socialist in ideology and a Capitalist in methodology".

Cloak of United Nations prestige and authority

Strong commissioned a report on the state of the planet, *Only One Earth: The Care and Maintenance of a Small Planet* for the first UN meeting on the environment, held in Stockholm in 1972 which he had been appointed to organise. He was then asked to establish a new UN body - the UN Environment Programme (UNEP) and became its first head. He later became Under Secretary General of the UN.

He correctly perceived that the impact of the prestige and authority afforded by the UN would be fortified if he could associate it with a socialist all encompassing agenda for global change. Strong's thrust was that the industrialised rich countries of the West had exploited poorer countries and must fund them to enable an equilibrium of participation in wealth to arise.

Strong understood that to prevail it was essential that the cause was vested with the prestige of an international authority. With de Gaulle he realised that *"Authority does not work without prestige, or prestige without distance".*

He also understood that the authority could not be a democratic entity subject to electoral oversight. He recognised, with Hannah Ahrendt,[64] that *"Bureaucracy is the form of government in which everybody is deprived of political freedom, of the power to act; for the rule by Nobody is not no-rule, and where all are equally powerless we have a tyranny without a tyrant."*

It was Strong's UNEP that co sponsored the new UN body – the 'Intergovernmental Panel on Climate Change' – in November 1988.

The IPCC was held out as being an impartial group of scientists to enquire into climate change. However it has no such remit. Its mandate is to *'provide scientific information relevant to understanding human-induced climate change, its natural, political, and economic impacts and risks, and possible response options'.* It was and remains exclusively concerned with establishing that mankind has been responsible for dangerous and abnormal global warming.

Moreover it was not at all impartial. Its first chairman was Professor Bert Bolin who had proposed to the first World Climate Conference at Geneva in 1979 the theory of dangerous human induced climate change and had reinforced it at the further meeting at Villach in 1985. The Chairman of the crucial 'Working Group 1' of the IPCC was to be Dr John Houghton an evangelical Christian, head of the UK Met Office and a close ally of Bolin - each an ardent believer in human induced climate change.

From the outset catastrophe and intimidation were the principal weapons of those who adopted global warming as their chosen prediction of doom. It was Houghton who stated:[65]-

"Unless we announce disasters no one will listen."

---

[64] Hannah Ahrendt 14th October 1906 – 4 December 1975) was a German-born American political theorist. Many of her books and articles have had a lasting influence on political theory and philosophy. Arendt is widely considered one of the most important political thinkers of the 20th century.

[65] John Houghton "Global Warming: The Complete Briefing". Lion Press 1994.

Houghton himself drafted the 'Summary for Policymakers' for his first IPCC 'report' of 1990. It predicted temperature would increase to end 2021 up to 0.5⁰C every 10 years. His models were thus predicting a rise every 10 years equivalent to 72% of the entire temperature increase over the previous 100 years.

The IPCC was used to launch the 'scientific' case for global warming by compiling 'reports' and convening conferences over the next 30 years at which were predicted ever more devastating climate events, using science to achieve political ends.

Overthrowing of energy policy. Political ends.

The political revolution was heralded at a so-called Earth Summit held at Rio de Janiero in June 1992.

It was the largest conference ever organised. It was attended by over 100 world leaders including Fidel Castro and 20,000 official delegates. Strong organised, and was responsible for, the entire content and process of the conference and procured that it should set up the Framework Convention on Climate Change. This was described as an international environmental treaty to combat "*dangerous interference with the climate system*", in part by stabilising greenhouse gas concentrations in the atmosphere.

The Rio conference led directly to the conference in Kyoto of 1997 which for the first time committed certain nations to severe reductions in CO_2 emissions. It was less than the believers were hoping for but it demonstrated the merging of politics and science activism to an extraordinary extent - and all within a time frame of less than 10 years. There followed the disaster of the Copenhagen conference in 2009 which achieved nothing. However again we were told we had just 10 years to save the Planet.

2015 Paris Climate Change Conference

In 2015 there was convened in Paris a fifth 'Conference of the Parties' - the name that had been given by Strong to the successive gatherings of nations which he initiated. There is another of these so called CoPs in Glasgow in November 2021. At each of these CoPs the latest IPCC gospel of doom is released as 'the science' and is composed for the purpose of inducing yet further assaults on our economic development.

Referring to the Paris CoP the following statement was made by the chief UNFCCC [66] organiser of that conference,[67]

> *'This is the first time in the history of mankind that we are setting ourselves the task of intentionally, within a defined period of time, to change the economic development model that has been reigning for at least 150 years since the industrial revolution.'*

It must also be understood that, in parallel with the environmentalist global warming doctrine, is the required transfer of wealth from western states to 'developing' countries.

The first environmentalist political conference at Rio (1992) provided that wealth transfer would be on the basis of a split between industrialised developed and developing countries. Included in these 'developing' countries - which were largely exempted from the proposals to cut CO_2 emissions - were China and India and so they remain. This was part of the Kyoto protocol.

In 2010 a Green Climate Fund was established to provide a reservoir of funds for draw down by 'developing' countries to be funded by developed countries which were parties to the Rio Convention. The 2015 Paris conference agreed that for the benefit of 'developing' countries those industrialised countries which were to make the greatest cuts of emissions would also contribute together an annual amount to this fund of $100,000,000,000 ($100 billion).

---

[66] United Nations Framework Convention on Climate Change.
[67] Statement of Christiana Figueres, Brussels February 2015.

Included among the beneficiaries of the GCF is China. It was one of the first to draw down – to the extent of $100,000,000 – in November 2019. It has not made any contribution to the GCF. In 2020 it commissioned 38.4GW of new coal plants. That is more than one large coal plant every week.[68]

Conclusion

Do not think that all this is some conspiracy theory that is being advanced in this paper.

I have no links with Big Oil. I knew nothing whatsoever about 'Climate Change' until I was persuaded to go to a meeting at Church House in 2009 which was intended as a debate between Professor Plimer and George Monbiot on the reality of global warming due to human activity. Professor Plimer is a most distinguished geologist who has written scholarly refutations of the theory of human induced climate change. George Monbiot is a weekly columnist of the *Guardian* newspaper. He is a leading proponent of modern environmentalism. Mr Monbiot did not appear to speak. He has recently stated,

> *"We've got to overthrow capitalism to stop climate breakdown. We've got to go straight to the heart of capitalism and overthrow it.[69]"*

These shocking statements are not just extreme Marxist outpourings by a journalist who has profited so much from the economy he wishes to destroy. They have always been and remain the settled intentions of diehard environmentalists.

Consensus

In essence the 'consensus' of the environmentalists is the assertion of a common belief. To express this it is said repeatedly that "The Science" is settled.

True science can never be 'settled'. It must be capable of falsification, the capacity for some proposition, statement, theory or hypothesis to be proven wrong. It is uncertain and only valid to the extent that it is not contradicted. The scientific method is exacting and scrupulous.

Counter 'consensus'

Historically it is untrue that there has been an overwhelming 'consensus'. After the Kyoto Treaty a petition was lodged by the Oregon Institute of Science and Medicine with 31,478 signatories. It urged the US government to reject the Kyoto Treaty and other similar proposals stating that limits on greenhouse gasses would hinder the advance of science and technology and damage the health and welfare of mankind. It continued

> *'There is no convincing scientific evidence that human release of carbon dioxide .. is causing or will in the foreseeable future cause catastrophic heating of the Earth's atmosphere and disruption of the Earth's climate.'*

There have been other such petitions and declarations[70] notably the European Climate Declaration[71]. As with all mass petitions there must be a discount for those who simply joined in for frivolous reasons but it is indisputable that many thousands had grounds for serious concern.

---

[68] San Francisco-based think tank Global Energy Monitor (GEM) and the independent organisation Centre for Research on Energy and Clean Air (CREA), 4th February 2021.

[69] George Monbiot Novara Media, 12th April 2019.

[70] e.g The Manhattan Declaration at the Non-Governmental Panel on Climate Change conference 2008.

[71] European Climate Declaration submission to UN Secretary General *by* Climate Intelligence Foundation by 500 signatory scientists *"[c]urrent climate policies pointlessly and grievously undermine the economic system, putting lives at risk in countries denied access to affordable, reliable electrical energy. We urge you to follow a climate policy based on sound science, realistic economics and genuine concern for those harmed by costly but unnecessary attempts at mitigation."*

Einstein himself asserted that *"that genius abhors consensus because when consensus is reached, thinking stops."* Max Planck, who with Einstein, revealed the secrets of modern physics, declared,[72]

> *"New scientific ideas never spring from a communal body, however organised, but rather from the head of an individually inspired researcher who struggles with his problems in lonely thought and unites all his thought on one single point which is his whole world for the moment"*

'Consensus' has been the basis on which over the ages valid and significant scientific observations have been dismissed and scientific advance and human understanding frustrated. There are many well known examples of this[73].

Maintenance of illusion of consensus

The environmentalist dogma dominating the IPCC requires that the illusion of consensus is maintained since it is used to justify the validity of its hypothesis of human induced global warming and the stifling of analysis, reason and debate. The maintaining of the 'consensus' is paramount. It is for this reason that the consensus falsifies when necessary in order to maintain political momentum.

> *We have got to ride this global warming issue. Even if the theory of global warming is wrong, we will be doing the right thing in terms of economic and environmental policy"* [74].

This willingness to distort and mislead is reviewed below in Section VII in the context of the 'conclusions' of the IPCC's 'reports'. But it operates no less in the efforts made to maintain the 'consensus' illusion.

Repeated false warnings as to predicted and fearful rises in temperature were uttered by Hansen, the principal director of GISS, to attract support for a supposed consensus . He used his authority as director of one of the two world custodians of surface temperature data records to lend its authority to his reckless assertions[75]. Those can now be seen as truly absurd. In 1986 Hansen predicted that the USA would heat by up by a further 2°C - 4°C - over a period of 14 years[76]. It increased over that period by 0.2% globally (Figure 11) and scarcely at all in the USA[77]. He persisted in wild predictions to Congress in 1988 all of which are contradicted by satellite evidence[78]. The IPCC used climate models for each of its 'reports' over the next 20 years which GISS manipulated to justify its hypothesis[79]. The models diverge very substantially from satellite evidence.[80]

The 97%

In 2010 there appeared a survey which purported to show that 97% of climate scientists agreed with the 'consensus' on global warming. It has been continually restated. In 2019 that claim was elevated to 100%[81]. Thus it is suggested that there is no scientist – with knowledge of the processes regulating temperature and content of the atmosphere including solar variation, Milankovitch cycles of orbital variations, the water vapour content of greenhouse gasses and the molecular radiative properties of

---

[72] Address on the 25th anniversary of the Kaiser-Wilhelm Gesellschaft, January 1936, as quoted in Surviving the Swastika: Scientific Research in Nazi Germany, 1993.

[73] e.g the Copernican theory as to rotation of the Earth suppressed by the religious consensus in the 17th century. In the 20th century Wegener's theory of plate tectonics and its explanation of subduction and continental shift was ridiculed by a consensus of scientists.

[74] "Paul Watson President of UN Foundation a private organization formed to support UN activities including propagation of climate change theory. Forbes 6th February 2013.

[75] The other being HadCrut in Essex.

[76] www.realclimatescience.com/2019/02/61-of-noaa-ushcn-adjusted-temperature-data-is-now-fake. and press articles cited.

[77] See Figure 7 for 1999 before it was tampered with in 2007.

[78] See Figures 10 and 11. His evidence to the US Senate Committee on Energy and Natural Resource in 1988 was that the 4 hottest years in 100 years recorded since the advent of recorded measurements had been in the 1980s rising to a peak in 1987.

[79] See Section VI which examines the tampering with data and graphs by GISS.

[80] See Figure 12.

[81] Powell, James Lawrence 20th November 2019. "Scientists Reach 100% Consensus on Anthropogenic Global Warming". Bulletin of Science, Technology & Society.

trace gasses – who considers that warming of the planet is attributable to any extent to natural occurring causes and who does not hold that it is exclusively the result of man made CO_2 emissions.

Such absurd 'findings' do not bear examination[82].

It was asserted that the 97% were 'climate scientists'. When the basis of the 97% was examined it emerged that the questions had been directed to "Earth scientists". There were 10,257 such persons included on the survey[83]. When these Earth scientists were approached it was decided that the disciplines of very many did not qualify them to answer, including physicists, geologists, astronomers, and experts on solar activity. Yet these are precisely the disciplines that most closely bear on issues of climate. Geology is the indispensable science for determining atmospheric conditions since the emergence of multicellular life 550ma (million years ago).

Nevertheless, having deemed them to be inappropriate, those responsible for the publication of the survey restricted it to 3,146 'Earth scientists'. Not satisfied with the percentages achieved with this more confined survey it was then limited to 79 respondents from the 'Earth science' community.

As would be expected, out of the 79 respondents 77 agreed that human activity had significantly contributed to warming. They divided 77 by 79 to get the 97%. Thus it was that of the 10,257 climate scientists approached the consensus was in reality confined to 0.75% of the total claimed. These few disciples however excluded all those specialising in specific areas of scientific enquiry that in the absence of bias would best qualify them to comment.

An attempt to breathe life again into the 97% was made in 2012 by an Australian psychologist and colleagues. It emerged that of the 11,944 papers reviewed only 65 stated specifically that the authors believed in human induced global warming.

As was explained in testimony to US Congress[84] by a lead author of the IPCC:-

> *"The 97% is essentially pulled from thin air; it is not based on any credible search whatever."*

Consensus is the language of politics. It can never justify the validity of an alleged scientific proposition. Moreover it is not the consensus among the believers that is to be feared. It is the consensus that has stupefied rational processes of enquiry among those who govern us.

Dogmatic suppression

> *'One of the most constant characteristics of beliefs is their intolerance. The stronger the belief, the greater its intolerance. Men dominated by a certitude cannot tolerate those who do not accept it.'*
> ***Le Bon***

Dogmatic suppression of opinion and doubt is the concomitant of belief in what is declared to be an incontrovertible truth. It is the watermark of religious intolerance and of anti-semitic hatred.

Environmentalism belief engenders the same quality of intolerance and even hatred though not to such degree or to such extremes[85]. Such an assertion may at first strike an impartial reader as an overstatement. However over the past 30 years there has accumulated such compelling evidence to justify it that some account of its vicious enmity is required.

Nigel Lawson (Lord Lawson[86]) published a mild but most persuasive book "An Appeal to Reason" in 1998. He did not reject the theory of global warming and simply argued that the science is far from

---

[82] For a good analysis see Professor D Easterbrook "Evidence Based Climate Science" 2nd edition p3, 4. Elesevier Inc.

[83] Enquiry revealed that the survey had been prepared by a Masters degree student at the University of Illinois.

[84] Dr Richard Toll testimony to US Congress: Full Committee Hearing – Examining the IPCC Process 29th May 2014.

[85] Charles Prince of Wales describes those casting doubt on the hypothesis as "headless chickens". Ed Davey Sec of State for Climate Change described them as "willfully ignorant".

[86] Chancellor of the Exchequer. Cabinet minister 1981 to 1989.

settled. He opposed the notion of scientific consensus claimed by the IPCC. He has spoken publicly of the enmity exhibited to him following publication of the book.

"I have never in my life experienced such extremes of personal hostility, vituperation and vilification"[87]

A sample of the nature of this bitter intolerance is the comment of George Monbiot,[88]

'Almost everywhere climate change denial now looks as stupid and as unacceptable as Holocaust denial.'

Those who question the dogma are described as 'climate change deniers' equating them by innuendo to neo-Nazis who deny the reality of death camps in Germany in 1942 – 1945. Giving warm endorsement to Monbiot's book on 'How to Stop the Planet Burning' the environmentalist Blog Gristmill concluded that,

'When we have finally gotten serious about global warming ....we should have war crimes trial for these bastards, some sort of climate Nuremberg'.[89]

Wikipedia

It was when undertaking research for this paper that I discovered the shocking bias of Wikipedia in its selection of articles on Climate Change. No information is provided as to the long and sincere scientific challenges to the environmentalist theory. It has an entire section devoted to "Climate Change Denial" with no balancing account of the contrary scientific case.

Editorial control of climate change issues passed years ago into the control of fervent climate change apostles. It was disclosed in 2009[90] that its senior editor – environmentalist William Connolley - had deleted over 500 articles and barred 2,000 contributors from its pages in a cleansing of heretical opinions. He created or re-constructed unique articles of Wikipedia on climate change.

Wikipedia still asserts that 'Nearly all actively publishing climate scientists (97–98%[3]) support the consensus on anthropogenic climate change' when it has been demonstrated that such claim is simply fanciful. Wikipedia, the world's most influential information source, excludes articles explaining the flaws in environmentalist global warming theory. It describes those who question its validity as guilty of *'denial, dismissal, or unwarranted doubt that contradicts the scientific consensus on climate change'* and describes the contra arguments as simply *'manufactured uncertainty'*.

It is to be deplored that balanced and empirical observations of data as to non-correlation of CO_2 levels and planetary heat, as to temperature variations over recent and geological periods and as to the negligible effect of additional atmospheric CO_2 should be denied to those seeking a balanced and comprehensive account of the existence, cause, and consequences of global warming.

It is surely fundamental that intense levels of scrutiny and review are required when vast – truly vast – amounts of national reserves of wealth are to be expended and when human lives are alleged to be at stake.

Is it not wicked that scientific expression should be so suppressed that it falls to private individuals to challenge the dogma and exhort those who direct affairs to arrest further descent into the mire.

---

[87] May 2014 Speech at the University of Bath.

[88] 21st September 2006. *The Guardian*.

[89] Grist: Environmental News and Commentary Grist Magazine, Inc.

[90] William Collins Journalist Canadian National Post cited in Christopher Booker Op Cit page 89 footnote 76.

BBC and the Press

Broadcasters and the Press compound the vilification of those who adduce evidence to challenge the gospel of climate change. Almost daily the BBC disseminates tales of hottest days on record, death by heat, bush fires, floods and the rest with no attempt to present balanced scientific enquiry. Typical of these assaults on free expression of established scientific knowledge is the comment of another Guardian columnist that *'climate change may be an issue as severe as war. It may be necessary to put democracy on hold for a while'.*[91]

The BBC has never attempted to give equal time and weight to dissemination of the flawed nature of IPCC 'reports' or to the basic contradictions in global warming theory. Conscious of accumulating public disquiet at its bias, in 2010 it commissioned an 'independent' report on 'The Impartiality and Accuracy of the BBC's coverage of Science' from Professor Steve Jones (a geneticist).

The flaws, misconceptions and mistakes in this exercise were so bizarre that they have been succinctly analysed in a book published by the Global Warming Policy Foundation[92]. But the appointment of Professor Jones was not of an independent scientist. He described those dissenting from the 'consensus' as a *'deluded minority'* whose views were similar to those who held that the 9/11 attack on the Twin Towers had been *'a US government plot'*.

Such is the deranging power of dogmatic belief that there appeared on the website of the Karl Franz University in Graz, Austria, the country's second-largest and oldest university, the following statement of Professor Richard Parncutt,

> *'As a result of that process some global warming deniers will never admit their mistake and as a result they will be executed. Perhaps it would be the only way to stop the rest of them. The death penalty would have to be justified in terms of the enormous number of saved future lives'.*

Conclusion

In response to questions made in good faith as to the validity of the global warming hypothesis its believers do not produce data to demonstrate its validity. They do not re-visit the formulations on which it is based. They do not attempt to falsify any contrary indications. What they do is to attack the persons who raise the questions. They seek to eliminate the ways in which their views can be ventilated. They vilify them and suppress all explanations of how climate varies, its causes and effects. They treat them as heretics.

As Einstein observed: *'Blind belief in authority is the greatest enemy of truth'.*

SUMMARY

Climate environmentalism uses science as dogma for political ends. It does so by promoting fear and guilt. The IPCC is a political bureaucracy. Authority is used to displace rational enquiry. Consensus cannot prevail over evidence nor does any consensus of climate scientists exist. Scepticism is the essence of the scientific method.

---

[91] James Lovelock May 2010, *Guardian*.

[92] Christopher Booker "A Study in Groupthink" pp 56,57 2018, The Global Warming Policy Foundation.

VI IPCC v. SCIENTIFIC METHOD

Nature of the Scientific Method

Strict scientific integrity is surely essential to sustain statements that are intended to dictate the overthrow of the energy basis of the modern age.

The IPCC must establish its hypothesis beyond verifiable scientific contradiction. That hypothesis is the existence of such peril for humanity from human induced global warming as can only be averted by eliminating all human emissions of CO_2. The burden of proof rests with the IPCC.

However uncertain may be the hypotheses and propositions of science it is imperative that they depend on rigorous scientific methods applied with unqualified integrity.

At the core of the scientific method is the process of eliminating errors of observation and conception by constantly subjecting propositions and theories to rigorous testing. That process requires the comprehension and examination of all the evidence that physics and observation provide. As physical evidence changes the concepts of science must change.

Richard Feynman

Professor Richard Feynman was the most distinguished physicist of the second half of the 20th century. Feynman expressed the scientific method as,

> *"the doubting that the lessons are all true....Science is the belief in the ignorance of experts."*

This is what he said of the indispensable requirement for the scientific method.[93]

> *"It's a kind of scientific integrity, a principle of scientific thought that corresponds to a kind of utter honesty—a kind of leaning over backwards. For example, if you're doing an experiment, you should report everything that you think might make it invalid—not only what you think is right about it: other causes that could possibly explain your results; and things you thought of that you've eliminated by some other experiment, and how they worked—to make sure the other fellow can tell they have been eliminated.*
>
> *Details that could throw doubt on your interpretation must be given, if you know them. You must do the best you can—if you know anything at all wrong, or possibly wrong—to explain it. If you make a theory, for example, and advertise it, or put it out, then you must also put down all the facts that disagree with it, as well as those that agree with it."*

IPCC. Evidence and Process

An examination of the origins and the practices of the IPCC reveals that is has no tenable claim to be a body governed by the scientific method. It is a political bureaucracy. Its judgments cannot be challenged by any democratic oversight. It is not an institute of science at all. It collates contributions from authors, few of them specialist meteorologists or paleoclimatologists, that support its declared founding purpose. That was and remains the justifying of a pre-determined dogma as to the existence of dangerous man made global warming.

The IPCC does not - and does not purport to - enquire into, much less to weigh in the balance, natural causes of temperature variation which both observation and basic physics suggest. It is a vehicle for

---

[93] Richard Feynman. Caltech Commencement Address, 14th June *1974*.

persuasion, conversion, and ingestion of belief. It fills the space left by our ignorance of scientific phenomena. It does so to intimidate and convert: not to reveal.

IPCC flaws as to evidence

Set out against each description of the following IPCC practices are the references to the Sections in this paper that justify them.

- It takes no account of verifiable analyses and calculations of the greenhouse effect made possible by a new field of physics[94] derived from quantum mechanics which demonstrate that increased densities of CO_2 have negligible effect on temperature. (Section II)

- It fails to give paramount regard to the evidence of satellites and balloon sonde measurements, where the greenhouse effect occurs in the atmosphere, which has been available from satellites since 1979. Such evidence displaces surface measurements and reconstructions of proxies by reason of precision, certainty, global daily coverage and lack of distorting factors (Section III).

- It gives no weight to the absence of any causal correlation of increased CO_2 density and rise in temperature both historical (400,000 years ago) and recent (2,500 years ago to modern times) (Section IV).

IPCC flaws as to process

The IPCC process involved in the producing of its 'reports' and particularly its *Summary for Policymakers* offends against the scientific method.

The main text of the 'report' is amassed from contributions from environmentalists of various disciplines and backgrounds many of whom have no claim to any scientific expertise and experience in paleoclimatology.

The texts of its 'working groups' are altered after first drafts if they conflict with the global warming dogma[95]. It produces a short document known as the *Summary for Policymakers*. It is a document which contains categorical statements as to the impact on human life of global warming. It is submitted to Government agencies each of which can insist on changes.

Those scientists who have contributed to the main text have no means for objecting to these political changes. Important and highly contentious declarations appear in the summary that are not in the text[96] and are often contrary to it[97]. Reliance is placed on distorted and selective data from unreliable sources[98] and unimpeachable evidence to the contrary is ignored.[99]

As we have seen[100] the founding moving force, Maurice Strong, was able to set up the IPCC using the United Nations to ensure both global dissemination of the consensus dogma and also to confer a borrowed prestige to its supposed authority. It has never been a vehicle for the scientific method.

Its process is flawed in three fundamental respects.

---

[94] Atomic, Molecular, and Optical Physics.

[95] See for examples Plimer Op cit pp 21,22. Booker C, *Global Warming A Study in Group Think*, 2018 The Global Warming Foundation IBSN 978-0-9931190-0-2.

[96] For the most recent example see graph 1 of the 2021 IPCC Summary examined below at Section VII.

[97] Evidence to House of Lords Select Committee *'The Economics of Climate Change'* published 5 July 2005.

[98] See for example Section below re Hockey Stick graphs and dataset sources for 2021 Summary for Policymakers.

[99] For the most recent example the reliance on defective PAGES2K sourced data see below Section VII despite incontrovertible satellite and balloon sonde evidence 1979 to date (see above Figure 11).

[100] Section V.

- It does not subject contributions to its reports to consistently applied rigorous review by climatologists and scientists in related disciplines including those who hold other or contradictory views. Publication is within a closed politicised system.

- It relies on modelling and reconstructed proxy evidence that is uncertain and prone to tampering and does not give precedence to observed data from satellite and balloon radiosondes.

- It distorts or conceals data when in conflict or inconsistent with its predicated constant upward trend of global warming or with its predicated cause of human CO_2 emissions.

Peer review

This is the process whereby the discipline of 'falsification'[101] is intended to be applied. But it is only an editorial tool. It is conducted at the editor's discretion. The process depends on the integrity of the editor. Yet it is the crucial stage at which the larger scientific community may refute, or put in context, or develop the theories and conclusions proposed for publication. The history of science shows that it is self-correcting over time and is only by this process can it be accorded more than temporary and provisional validity.

Scientific journals and papers should take no position on any issue. They should be indifferent to any political implication. They should publish 'the spectrum of competing hypotheses'[102] and defer to the scientific community the process of validation or replication over time.

The environmentalists at the IPCC use peer review not to expose their hypothesis of global warming and its alleged impact to critical review but to create the impression of common scientific acceptance of the theory. By this means it claims a supposed impartiality. The intention is to convey that only papers which have been subjected to peer reviews by other independent experts would be relied on. But the reality is otherwise.

Abuse of peer review process

There are many examples of how far the IPCC has fallen short of this process of validation or falsification. In particular it has not admitted papers from scientists of disciplines that directly bear on climate which reject the consensus. Nor does it publish countervailing papers on natural causes of climate change which have been its causes throughout the Earth's existence.

The following examples are taken from the content of the 2007 IPCC 'report'. This was the 'report' to be used for the 2009 Copenhagen conference. It was followed by the 'report' for the 2015 Paris conference.

One of the key chapters of the 2007 IPCC 'report' was that concerning understanding and attributing climate change. It claimed that it had been approved by 1,500 'climate scientists'. The IPCC claim was false. A study into the claim revealed that it had been written by 53 authors. 60% of these came from research units in the USA and in Britain that were emphatically committed to the 'consensus' on global warming and its principal advocates. Most of them had co-authored papers with each other or had 'favourably reviewed' each other's work.[103]

It also emerged that the much of the content of the 2007 IPCC 'report' as to the 'impacts of climate change' had not been based on any peer review whatsoever. The Working Group II had not drawn on any peer reviewed literature but had derived content from propaganda material put out by climate activists and environmentalist pressure groups.

---

[101] Karl Popper holds that all scientific knowledge is provisional. It must also be falsifiable. Falsifiability is the assertion that for any hypothesis to have credence, it must be inherently disprovable before it can become accepted as a scientific hypothesis or theory.
[102] The analysis of competing hypotheses is a methodology for evaluating multiple competing hypotheses for observed data. It was developed by Richards J. Heuer, Jr., of the Central Intelligence Agency in the 1970s.
[103] Christopher Booker Op Cit 2018 p 43 citing "Prejudiced Authors Prejudiced Finding" Science and Public Policy Institute John McLean 2008.

The 'Glaciergate' scandal arose from the prediction of the disappearance of Himalayan glaciers. This was exposed when it was shown that this assertion was made in a local interview in a small environmental magazine in 1999 but which was quoted in 2005 by the WWF[104]. It was dismissed by a lead IPCC author as not even worth discussion. It was still published.

IPCC claimed that global warming could well destroy 40% of the Amazon rain forest. This rested on a propaganda leaflet put out in 1999 by a Brazilian environmental activist group. But that prediction was not made on the grounds of the impact of global warming but because of logging and fires.

A further IPCC claim was that droughts caused by global warming were likely to result in falls of 50% of African crop yields. This was traced to a single paper by a Moroccan academic who himself claimed it was based on reports to three North African governments. Not one of these reports made any such claim. They had forecast that crop yields might actually rise.

These claims – Himalayan glaciers, Amazonian rainforest and drought in Africa – caused great public concern. This resulted in a comprehensive investigation[105] as to the sources of statements in the 2007 IPCC report. It revealed that nearly one third of all scientific reference had not been peer-reviewed academic studies. They were *'newspaper and magazine articles, discussion papers, MA or PhD theses , working papers and advocacy literature put out by environmentalist groups'*

It was said of John Maynard Keynes that he once declared *"When the facts change I change my mind. What do you do?"*

Process of modelling

Climate modelling is a component of climate science. It is the basis of predictions as to future change. It can also help with the causes and extent of past changes. But it cannot be preferred to direct measurements. Nor can it ever be free from uncertainty.

The basis of modelling is a notional three dimensional grid consisting of grid boxes and sub-grid boxes above the surface of the Earth. There are about one million for atmosphere and 100 million for the ocean. Physics can be applied to calculate horizontal and vertical movements between grid boxes of air, water and energy. This process has to be repeated countless millions of times to determined simulations of climate for 100 years – ten minutes requires over five million times.

But there are serious variables that affect conclusions[106].

Changes and differences of conditions actually occur on a much smaller scale than that of a grid box or the subdivided sub-grid box. This factor requires assumptions to be made. Since modellers make their own assumptions the results as between models will vary widely – this can be seen from coloured graphs of combined models included in this paper (Figure 12 Section III). An extensive exercise of judgement is required. In particular fluctuations in height and extent of clouds can have as much impact on solar radiated energy as human influences. Increases of precision are not possible with the current state of technology.

Moreover the relatively small depth of the oceans and height of the atmosphere require very thin grid boxes for any accurate description of vertical variation. At the surface of the Earth and upwards in the atmosphere for about six miles there are turbulent movements of energy and water vapour.

[104] World Wildlife Fund for Nature.

[105] Christopher Booker Op Cit 2018 p 44 Citing survey report of Donna Laframboise.

[106] See for a clear and accessible account of modelling see Chapter 4 Professor Koonin S.E "Unsettled. What Climate Science Tells Us, What It Doesn't and Why it Matters" BenBella Books Inc 2021 from which this part of the paper is derived.

Evaporation causes flows of radiated energy from the oceans. This is over 30 times that of human influences. Assumptions as to this phenomenon are critical to any modelling as to these movements.

Another factor of uncertainty is that for accuracy and precision it would be necessary to establish the initial detailed states of the ocean and atmosphere as the start of the time span. But this is not possible with current means and systems of observation. As explained by Professor Koonin [107] even if it were indeed possible the level of chaos would render most of the details irrelevant after a few weeks. Thus only the broad features can be correctly established (jet stream, major ocean currents).

Tuning

A further variable element of climate modelling is the process of 'tuning'. This, put very simply, is the adjustment of the model to achieve an improved balance between observed conditions and the physics parameters set for the sub-grids. But such tuning is often used as a means of producing a pre-determined result. It is in any case impossible to tune the parameters to match all observed properties of the climate system[108].

'Feedbacks' are a further critical component of models. CO_2 concentrations will either diminish or amplify their warming influence. Thus as the planet warms up snow and ice will diminish so decreasing albedo[109]. But a less reflective Earth absorbs more solar radiation and increases water vapour which enhances heat interception with high cloud and reflectivity with low clouds. These factors have to be allowed for in tuning and each model will give different conclusions.

It has been well said that *"Tuning may be seen indeed as an unspeakable way to compensate for model errors".*[110]

Limitations of modelling

All of these factors must be borne in mind when discussing climate models and none more so than the Couple Model Intercomparison Project, particularly as it relates to warming since 1979 (CMIP5 and CMIP6) when satellite data was available.

As mentioned above the nature and limitations of climate modelling require that much content has to be assumed. Ex post facto adjustments have to be made for inconsistencies between set parameters and actual observations. This is not to discredit those compiling climate models. These adjustments and assumptions are inherent in the modelling process as it is currently possible within the present computer power and capacities.

A recent example[111] is referred to by Professor Koonin. It relates to UK climate scientists adjusting snow cover and dimethyl sulfide impacts so altering the albedo factor (reflectivity of Earth). He also cites a tuning of a well known model of the Max Planck Institute by a factor of 10 for the reason that the value originally adopted resulted in twice as much warming as had been observed.[112]

With the limitations – or the opportunities, if so considered – of the modelling process considerable distortions may result from extrapolations of imperfect data. They may equally be induced by setting parameters weighted as to climate influences that will indicate a trend. The process can easily be subject to abuse.

---

[107] Koonin Op Cit Chapter 4.
[108] Koonin Op cit p 84 – 86.
[109] non-dimensional, unitless quantity that indicates how well a surface reflects solar energy.
[110] Bulletin of the American Metrological Society 98 (2019) 589 – 602 cited in Koonin Op cit p 86.
[111] See for example Sellar et al "UKESM 1 (United Kingdom Earth System Model) Description and Evaluation of the UK " Journal of Advances in Modelling Earth Systems II 2019 4513 – 4558.
[112] Mauritzen et al "Developments in the MPI – M Earth System Model Version 1.2 and its response to Increasing CO2" Journal in Advances in Modelling Earth Systems 11 (2019) 998 – 1038 cited in Koonin Op Cit p 86.

The imperfections of such modelling are illustrated in the ensemble of models used in the IPCC reports of 2013 (for the 2015 Paris conference) and 2021 (for the 2021 Glasgow conference).

The IPCC 2013 'report' was based on climate model CMIP 5. The divergence of these models from gilt edged satellite and balloon sonde measurements has been discussed above (see Figures 11 and 12 and text). As is shown (Figure 12) CMIP5 was comprehensively compromised by these incontrovertible data sources. The latest IPCC 'report' is based on CMIP6 (2019/20) and so significant are the distortions of these models that they are the subject of Section VII.

Divergence from satellite data

However there is an important further issue. Not only do CMIP models give graphs of climate change diverging widely from satellite data. They also diverge from each other. This is illustrated by the visual impact of these models graphs of Figure 12. The simulated global average temperature varies as between models by approximately 3°C . Yet the observed increase in warming over the period of over 100 years prior to these models was approximately. 1°C . So the variation is greater than the totality that they are purporting to illustrate and explain.

Moreover the CMIP5 ensemble (2008) used to underwrite the IPCC 2014 report shows an even wider overall divergence of the component models than the CMIP3 (2005/6) used to underwrite the conclusions of IPCC 2007 report. As Professor Koonin points out the later version was actually more uncertain than the earlier one.[113]

What crucially vitiates these models, however, is the failure to depict, much less to explain, the sharp and defined warming period from 1910 to 1940 when CO_2 emissions had not seen much acceleration (Section IV). That warming period is similar to - and in the United States greater in intensity than - the warming period of 1975 -1998.

These models do not depict actual known observed measurements of global temperature for the second half of the 20th century with sufficient precision and consistency to be at all useful. They cannot even produce the past.

Climate sensitivity

This is a technical climatespeak term. It means the global temperature rise following a doubling of CO_2 concentration in the atmosphere compared to the pre-industrial level of 280 ppm (1850).

Climate sensitivity is a critical component of climate assessments. The higher the estimated climate sensitivity the larger the predicted temperature increase.

Cooling is effected by radiation from the surface of the Earth. Cooling is reduced by 12.1% by greenhouse gasses, ignoring CO_2. CO_2 adds a further 7.6% reduction in cooling power at 400 ppm. But as has been explained in Section II the doubling of CO_2 concentrations account for only a further 0.8% reduction in cooling power[114] which diminishes with increased cloud cover.

Yet the CMIP 6 ensemble models provide for climate sensitivity to a far greater extent than even the IPCC 2014 report model. It is now in the range 1.8°C and 5.6°C. To justify this an attempt has been made to ascribe such predictions to cloud interaction with aerosols[115] (fine atmospheric particles) which induce the formation of reflective clouds. Aerosols interact both directly and indirectly with the Earth's radiation budget and climate. As a direct effect, the aerosols scatter sunlight directly back into

---

[113] Koonin Op cit Ch 4.

[114] Koonin Op cit p 53 and Easterbrook pp 321, 322. See also *Relative Potency of Greenhouse Molecules*. Professor W A Wijngaarden Dept Physics York University Canada and Professor W. Happer Dept Physics Princeton University USA. Koonin Op Cit p 54.

[115] Minute particles suspended in the atmosphere including from incomplete combustion of fossil fuels and volcanic eruptions. When sufficiently large they scatter and absorb sunlight. Scattering of sunlight can reduce visibility (haze) and redden sunrises and sunsets.

space. As an indirect effect, aerosols in the lower atmosphere can modify the size of cloud particles, changing how the clouds reflect and absorb sunlight, thereby affecting the Earth's energy budget.

However, as explained by the National Center for Atmospheric Research, the effect of aerosols on cloud formation is simply speculative[116].

The significance of all this is that IPCC predictions of global temperature use far wider ranges of sensitivity than can conceivably be justified. The use of such exaggerated multiples is reckless given the consequences that will result if they are relied upon. The IPCC rely on models to underwrite its forecasts of catastrophe. Its modelling is palpably defective for the reasons set out above.

Homogenising surface temperature data

We have seen (Section III) how it is that satellite and balloon radiosondes have since 1979 provided by far the most representative and reliable data for the temperature of the Earth.

As explained (Section III) satellites measuring temperature monitor the entire planet. They do so each day. Balloons measure temperature at all levels into the upper air. The upper air is much more horizontally coherent than at the surface. However surface temperature stations have the serious imperfections of being scattered over a small area relative to the Earth's surface, do not cover more than an inadequate percentage of land and oceans and are contaminated by proximity to urban heat effects and frequent site and instrument changes.

Surface thermometer temperature records over the period of 100 years prior to 1979 remain as the best available data source. What renders these surface measurements open to doubt as to their historical validity is the process known as homogenisation. This is intended to remove spurious factors which have nothing whatever to do with climate or temperature.

However it is a process of adjustment that has been applied to sustain the IPCC hypothesis of a warming trend in step with the increase in atmospheric density of CO_2.

To show a consistent warming trend married to a consistent CO_2 rise would conflict with the pronounced rise of actual global warming of 1910 to 1940 and the steep global cooling from 1940 to 1975. These negate the IPCC hypothesis. Accordingly from 2000 GISS "cleansed" historical data to produce the steady warming necessary to establish a false correlation with CO_2 increase[117].

Figure 23

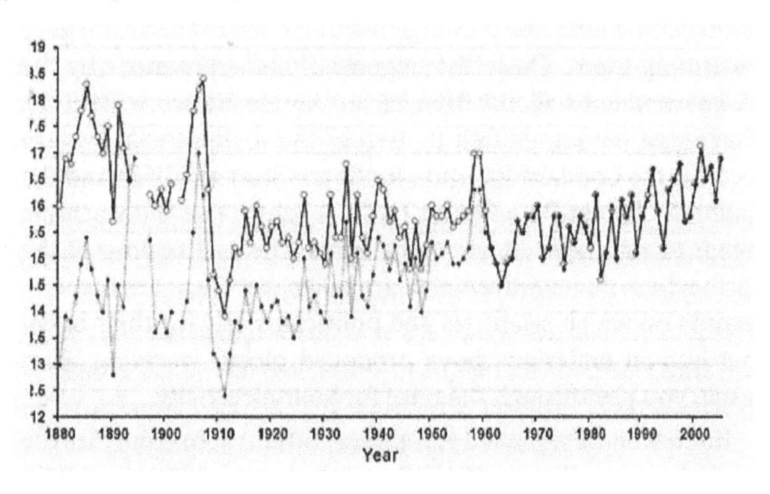

[116] *"Cloud aerosol interactions or on the bleeding edge of our comprehension of how the climate system works and it's a challenge to model what we don't understand. These modellers are pushing the boundaries of human understanding and I am hopeful that this uncertainty will motivate new science"* National Center for Atmospheric Research "Increased Warming in Latest Generation of Climate Models Likely Caused by Clouds" Phys.org June 2020. Cited by Stephen Koonin Op cit pp 93 and 266.

[117] Professor Ian Plimer Emeritus Professor of Earth Sciences University of Melbourne *"Climate Change Delusion and the Great Electricity Rip Off"* 2017 Connor Court Publishing Pty pp 270 – 272.

This graph illustrates the impact of homogenising the original temperature record (thick line). The effect of these adjustments are not the minor changes contemplated for valid homogenisation. The intention and the effect is to show an average temperature increase since 1880 consistent with the 'consensus' but contradicted by reality. The weather station data on the previous graph is taken from what is now an urban measuring station (Davis Ca USA). GISS simply adjusted the historical data downwards. It is to be noted that in 1880 Davis was a rural weather station and urban heat island adjustments thus do not justify such tampering with the data.

These homogenous adjustments occur routinely in official records kept by bodies that promote the consensus dogma. The following are examples out of many perpetrated by the Australian Board of Meteorology and the New Zealand National Institute of Water and Atmosphere[118].

Figure 24

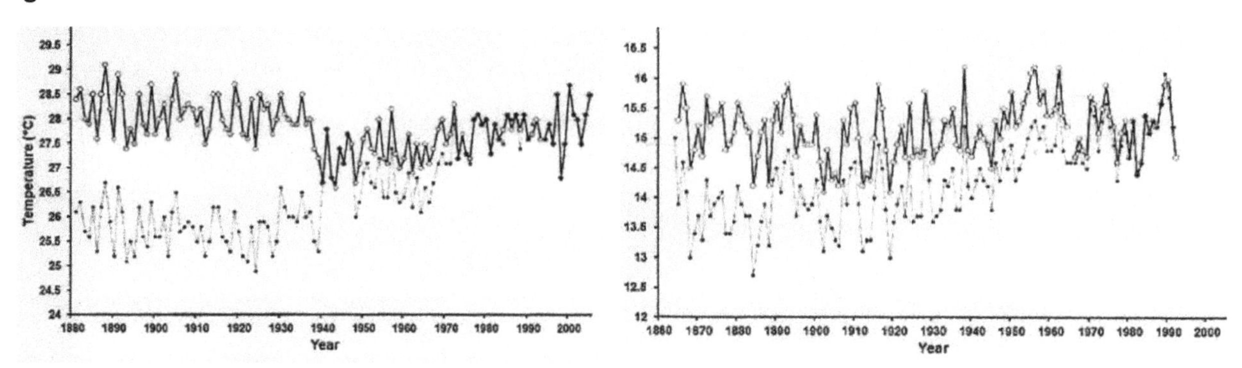

The thin line with black dots is the homogenised data trend. The thick line with circles is the raw original data. The left hand graph is taken from the Darwin station. The right hand graph is from the Auckland station. Each of these adjustments is made to show a warming trend.

Such adjustments are not always accounted for by a temptation to show an average gradual and constant rise in global warming since 1900. The adjustments however do show this trend which would not otherwise appear. They serve to fit the record with the dogma of consensus.

This created disquiet and the process was discredited when satellite and balloon sonde date showed that observational data differed greatly from climate models based on other temperature data.

Willful Distortion

Removal of Medieval Warming. The first 'Hockey Stick' IPCC 2001

No one sought to challenge the occurrence of the Medieval Warming period of 950 AD to 1350 AD until it became clear that the 'consensus' as to global warming was weak.

In 1997 a survey of US State Climatologists found 90% of them agreeing that *"scientific evidence indicates variations in global temperature are likely to be naturally occurring and cyclical over very long periods of time"*. The IPCC's' first report (1990) had accepted the evidence that at times in recent historical past temperatures had been higher than at the end of the 20th century. That report included a graph of climate change over the last 1,000 years – it is reproduced in Figure 25.

Thus to sustain the hypothesis of global warming in linear correlation with rising CO_2 and to fortify the consensus it was necessary to dispose of the Mediaeval Warming Period. Accordingly the IPCC adopted wilful distortions of climate data to produce a false prediction of rocketing temperature rise in close alignment with the ascent of CO_2 in the second half of the 20th century.

---

[118] Professor Ian Plimer Op Cit 2017 Connor Court Publishing Pty pp 272- 274.

Professor Jonathan Overpeck was one of a small group working closely with the IPCC. He was in Working Group 1 which prepared the Summary for Policymakers. He was later Coordinating Lead Author for the IPCC 4th Assessment (2007). In an e-mail to Professor Deming[119] in 1995 Overpeck had stated that *"we have to get rid of the Medieval Warm Period"*. That warming period disrupted utterly the consensus hypothesis of steady climb of temperature in line with increasing density of atmospheric CO_2.

There was also a further reason for the concern. This was the undeniable fact (see Figures 19 and 21) of the modern warming of the planet from 1910 to 1940. It was the highest since the 13th century. There was also the 30 years cooling from the mid-1940s. Each of these was in conflict with the unbroken rise of CO_2 in line with fossil fuel emissions.

As set out in Section IV the period from 950AD to 1300AD saw surface temperature rise to a level at up to 2°C higher than today. The first IPCC 'report' accepted this and included a graph showing temperature variations over the past 1000 years.[120]. The Medieval Warming itself had been preceded by the Roman Warming of 250 – 450 AD with temperatures 2 – 3°C warmer than now.

Figure 25

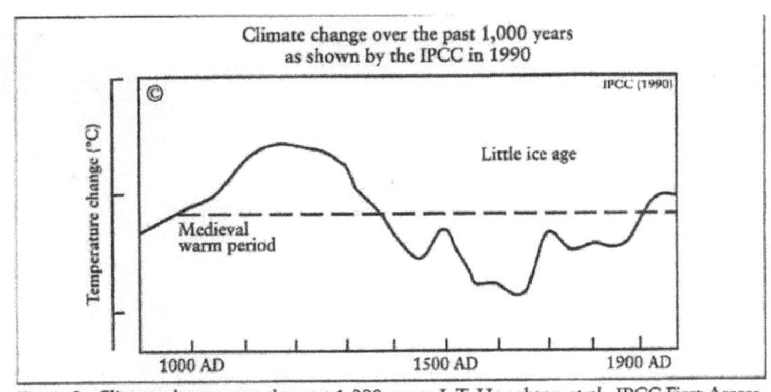

Figure 2: Climate change over the past 1,000 years: J. T. Houghton *et al.*, IPCC First Assessment Report, 1990.

Three years after the Overpeck e-mail (1998) the leading science journal 'Nature' published a graph prepared by a PhD 'researcher' Michael Mann. It simply ignored these historic fluctuations in recorded temperature and purported to show a sudden and very steep acceleration in temperature over recent years to a level higher than for a 1,000 years. This gave it the appearance of a hockey stick.

Figure 26

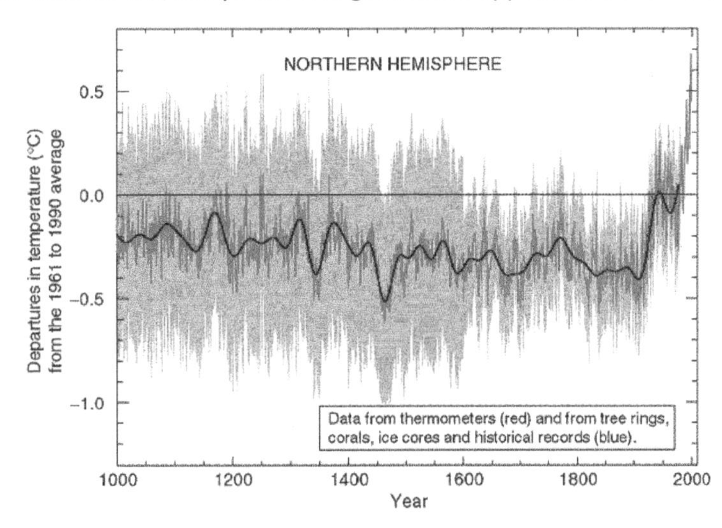

The four centuries of the Little Ice Age which ceased in about 1850 and the entire Medieval Warming period were eliminated.

---

[119] A geoscientist at the University of Oklahoma.

[120] IPCC First Assessment Report 1990. J T Houghton et al.

The hockey stick graph was immediately adopted by the IPCC. It formed the first page of the key *"Summary for Policymakers"* of the 2001 'report'. It was repeatedly relied on in the text of the 'report' and a vastly enlarged depiction of it was showcased at the 'report's launch. A temperature rise of no less than 5.8^{0}C by the end of this century was deemed to be probable by one of its model programmes – 2^{0}C higher than the 1996 'report' estimate 4 years earlier or over 8 times temperature recorded increase for the entire 20th century (0.7%).

When the flawed evidence of bristlecone pine rings was removed the Medieval Warming period re-appeared. It was then discovered than Mann's algorithm had given 390 times more emphasis to Californian bristlecones – which had anomalous hockey stick patterns – than to the trees from Arkansas showing a much flatter line. Moreover the algorithm had been programmed to mine for hockey stick shapes. It did so whatever data was introduced – of any kind.

Two Congressional inquiries as to Mann's graph were held and it is now utterly discredited. It had never been subjected to peer review.

Tampering with temperature records: HadCRUT4 and GISS

This distortion of temperature records underlay the appalling disclosures known as 'Climategate'.

E-mails from the database of the Climate Research Unit of University of East Anglia – one of the key providers of data and information to the IPCC - came to light. The disclosure of these exposed a deliberate intention to remove the blip of the 2Oth century oscillations.

The 'Climategate' officials wanted to get rid of the 1940s cooling blip. As was said in one e-mail[121],

"So if we could reduce the ocean blip by, say, 0.15deg^{0}C then this would be significant for the global mean – but we'd still have the land blip. It would be good if we could get rid of the 1940s land blip, but we are still left with "why the blip"[122].

Included in this revelation of a cynical attempted conspiracy to mislead was Professor Jones of the Climate Research Unit attached to the Hadley Meteorological Office. Jones proposed to colleagues that they should delete compromising e-mails.

The sharp divergence of the surface temperature records of HadCRUT and GISS – the official custodians of such records – from the far more reliable satellite and balloon sonde data was so significant that it led to an investigation in 2015 into the surface records put out by HadCRUT[123] originally in 2012 but updated from time to time.

Satellite records were showing 1998 as much warmer than any subsequent year. This spike was due to the El Niño event of that year.

The Climategate e-mails exchanged between leading consensus climate scientists had discussed how to "adjust" observed temperature trends to better support their theory by dealing with the problem of the sharp 1998 spike and decline in temperature since 1998 so as to replace them with a gradual looking rise. It was discovered that the surface records of HadCRUT had been adjusted to create this very impression. The Had CRUT3 report had shown 1998 as 0.007^{0}C warmer than 2010. But its subsequent version HadCRUT4 revealed that the 1998 temperature spike had been reduced and the figure for 2010 had been adjusted upwards[124].

---

[121] E-mail 27 Sept 2009 Tom Wigley to Phil Jones [director of CRU] cc Ben Santer.

[122] E-mail Tom Wigley of the University Corporation for Atmospheric Research wigley@ucar.edu to Phil Jones p.jones@uea.ac.uk. Subject: 1940s Date: Sun 27 Sep 2009 23:25:38 cc Ben Sinter santer1@llln.gov.

[123] UK Met Office Hadley Centre and the Climate Research Unit under Professor Jones.

[124] See the succinct analysis of all this by Christopher Booker Op cit pages 58 – 60.

The HadCRUT 4 was itself based on data provided by GISS[125]. It claimed that 2014 was warmer even than the sharply spiked temperature of 1998. This was false. It was contradicted by satellite data (see Figure 11). GISS had altered records of what purported to be a scientific record. This enabled the 'consensus' to continue to propagate its dogma of inexorable rise of temperature in tandem with rise in CO_2 and with the global warming theory.

It emerged that GISS had relied on evidence so flimsy as to undermine altogether any conclusions derived from it. It identified sharp temperature rises in South America from Brazil to Paraguay. Paraguay had just three weather stations. GISS nevertheless reported for each of them an increase in temperature between 1950 and 2014 of 1.5^{0}C. That is more than twice the 0.7^{0}C increase for the entire 20th century – as to which there is no dispute. But archived data originally recorded in those decades had shown a **cooling** trend of 1^{0}C.

GISS had again simply adjusted earlier temperature data downwards and recent temperature upwards.

Similar distortions of data were made for Arctic data. Investigation of the GISS record of the Arctic circle revealed data from weather stations between 52^{0} West (Canada) and 87^{0} East (Siberia). In every case GISS had adjusted older data downwards and later data upwards. Some of these adjustments were of more than 1^{0}C. The Reykjavik measured temperatures of the 1930s and 1940s were corrupted but nothing was subtracted from post 1980.

Figure 27

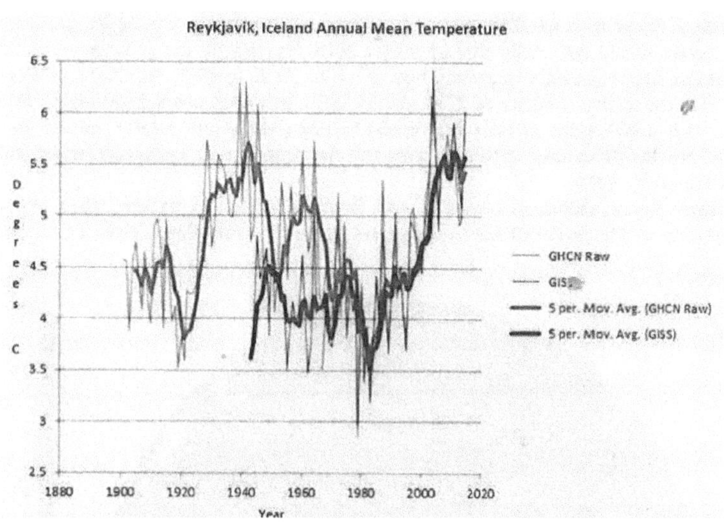

The same tampering was disclosed by other researchers showing that it had occurred in data recording across the world. Data from specific weather stations in Australia had been converted from an 80 year cooling trend of 1^{0}C per 100 years to warming trend of 2.3^{0}C. The absence of any warming trend for New Zealand from 1850 to 1998 was converted into a warming trend per century of 0.09^{0}C.

The manipulation of temperature data for the United States by GISS was disclosed in 2007[126]. The hottest temperature on record for the United States had been 1934. But GISS altered data for the US Historical Climatology Network[127] by downgrading the original recorded temperature figures for the 1930s and increasing those for recent years. This replaced 1934 with 1998 as the hottest year in American history contrary to the contemporary records. The 1930s were the years of the 'dustbowls' and searing heat.

More recently this tampering with data was shown to have been applied to sea level records. Main records of sea levels are maintained by the PSMSL[128]. Data records for the Indian Ocean had been subjected to significant alterations to give the spurious impression that sea levels recorded as stable

[125] Goddard Institute for Space Studies.
[126] Stephen MacIntyre: Climate Audit.
[127] A dataset used to quantify temperature changes in the contiguous USA adjusted for factors that bias temperature trends.
[128] Permanent Service for Mean Sea Levels (PSMSL).

or falling had actually been rising sharply. Lowering for earlier years and increasing for later years was the method adopted to achieve this.[129]

Any one of these shocking revelations should have fatally impaired the purported status of the IPCC's guidance to 'Policymakers' as being derived exclusively from scrupulous peer-reviewed and falsifiable propositions of independent scientists concerned with the causes of alleged man induced global warming.

The 2021 IPCC Hockey Stick

However the IPCC has, astonishingly, resurrected a further hockey stick in its latest 2021 'report' based on similarly distorted evidence as to temperature[130]. This time the hockey stick shows a prodigious leap from a dull and constant alleged historical record.

Page SPM 7 of the Summary for Policymakers of the 2021 IPCC 'report' sets out two graphs. One shows a purported a slight gradual decline of global temperature over 1000 years with a vertical rise over the last 60 years (Panel a) and the other (Panel b) shows purported very sharp increase in temperature for those 60 years corresponding to the CMIP6 climate model simulations.

The text of the IPCC graphs makes some conclusions that are the stuff of fantasy. It claims that temperature is now higher than in the past 100,000 years. It shows an increase of 1.5°C from 1910[131] to 2020. It claims that rocketing of temperatures has occurred over the last 60 years. It eliminates the Roman Warming, the Medieval Warming, the Little Ice Age, the sharp rise of the 1920/30s and the 1940 - 1975 severe fall.

The final Section of this paper is devoted to an analysis of the basis of these IPCC graphs and of the IPCC statements.

SUMMARY

The IPCC abuses the scientific method. It was founded to create belief in dangerous global warming and procure the overturning of our energy base. It denies solar cycles and ocean events as drivers of climate change. Its models are displaced by satellite observational evidence. It knowingly uses misleading graphs and distortions of data. It fabricates data to conform to its dogma.

---

[129] Dr Albert Parker and Dr Clifford Ollier "Are the sea levels stable at Aden, Yemen" 2017: 1:18. Cited in Booker Op Cit p 60 footnote 84.
[130] See Steven MacIntyre Climate Audit August 2021 *"The IPCC AR6 Hockeystick"*.
[131] Note: the graph only calibrates less than 0.5 variations nor does it calibrate the selected years.

VII THE IPCC REPORT AUGUST 2021: SUMMARY FOR POLICYMAKERS

The 2021 'report' rehearses the claim that the rise in temperature over the 20th century as the planet emerged from the Little Ice Age was without precedent. It relies for its impact on two leading graphs that appear only in the Summary for Policymakers and not in the text. These are reproduced below.

Figure 28

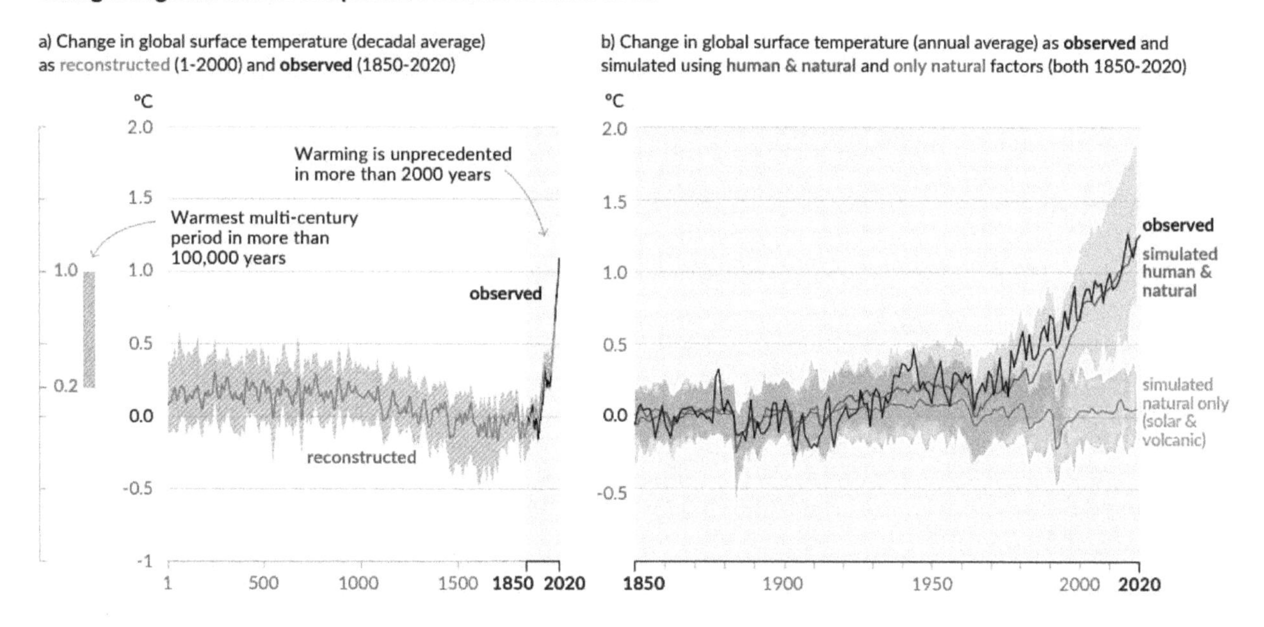

The Summary for Policymakers is the document that is directed to governments and their agencies. It is the one which is relied on. It is doubtful that the opaque text of the 'report' is reviewed in any detail, if at all, by Ministers.

The IPCC text under these graphs is as follows

"The vertical bar on the left shows the estimated temperature (very likely range) during the warmest multi-century period in at least the last 100,000 years, which occurred around 6500 years ago during the current interglacial period (Holocene). The Last Interglacial, around 125,000 years ago, is the next most recent candidate for a period of higher temperature."

It is not clear if the text intends to exclude the current date when it speaks of "the last 100,000 years" as if implying it is higher now. However that may be the left hand graph shows that the maximum temperature of the Holocene period 6,500 years ago as being actually **lower** than the temperature in August 2021.

The 'report's findings depend on the following misrepresentations.

1. That average global temperature has been in gradual decline from 1000 AD until 1850 with only slight fluctuations of 0.3°C but with a rise since then of 1.5°C

2. That warming as of August 2021 is unprecedented, even exceeds the warmest period of this interglacial Holocene period and is only exceeded by the peak of the preceding interglacial period 125,000 years ago.

These are not what are known to lawyers as "innocent misrepresentations". They are, in technical legal terms, 'fraudulent' misrepresentations in that they have been made either with knowledge of the true scientific position, or reckless as to what that might be, with the intent that they are relied on and are relied on to the detriment of those to whom they were made.

Elimination of Roman and Medieval Warmings

What these graphs show is the re-emergence of the utterly discredited Mann 'hockey stick' format. This is reproduced again below for convenient comparison with the IPCC 2021 graphs. The IPCC graph (left hand) relates to a span of 2000 years. The 2001 Mann graph uses 1,000 years.

Figure 29

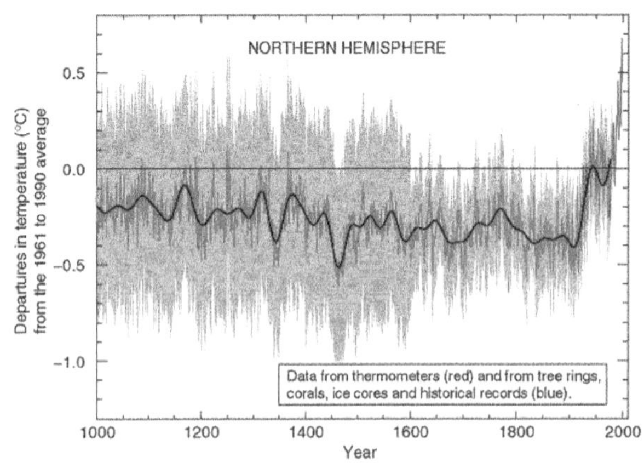

The juxtaposition of the Mann 1998 fabrication (Figure 28) and the IPCC left hand graph (Figure 27) gives the lie to the purpose of these false depictions of the trend and severity of climate variation.

The Mann graph has been described as fraudulent by leading climate scientists in that it knowingly manipulated data in order to eliminate the Medieval Warming period. A libel action brought by Mann for this allegation was dismissed 'with prejudice'[132].

We have seen (Section VI Figures 25, 26) that Mann had grossly exaggerated and distorted tree ring temperature 'proxy' data and used an algorithm programmed to produce the sudden curve at the end of the hockey stick – its blade. It was the lead graph of the IPCC 2001 report. The 'Climategate' scandal revealed e-mail exchanges with Mann (and others) which illuminated how data had been radically truncated by Mann in order to scrub out the record of warming and cooling and produce almost level average temperature record over the 1,000 years before a leap in the third quarter of the 20th century.

Twenty years on and the left hand IPCC 2021 graph also shows an average level then gradual decline in temperature over 1975 years followed by a vertiginous vertical rise from around 1975. This graph has inadequate calibrations[133] but the trends are clear.

The left hand graph (from 1 AD) removes the rise and peak of temperature in the Medieval Warming period. It also removes the peak of the Roman Warming period. They are simply eliminated. Nor are the steep declines of the Little Ice Age recorded.

The right hand graph (from 1850 AD) smothers the peak of the 1920s/1930s warming. It is simply not there. That peak was not matched until the freak year of 1998, itself due to the El Niño spike. The severe cooling of the 30 years from 1945 is not there. It has been cynically ignored.

The effect and intention of these deliberate truncations is to show 'Policymakers' that temperature has been in modest decline over centuries ending with a terrifying rise just 45 years ago.

---

[132] Dr. Timothy Ball environmental consultant former climatology professor at the University of Winnipeg, Manitoba, a having a doctorate in climatology from the University of London, Queen Mary College, London. He is Chief Science Advisor of the International Climate Science Coalition and a Policy Advisor to The Heartland Institute. He commented that *Michael Mann should be in the State Pen not Penn State*".

[133] It uses 500 year spans of time and 0.5°C gradations for temperature. The Mann hockey stick graph was easier to follow as it was calibrated in 100 year cycles. The 2021 is far more generalist even than Mann.

The acceleration of a supposed upward trend of temperature from 1980 is in direct contradiction of the accurate and comprehensive data provided by satellite and balloon sondes (Figure 11). Average temperature rise since 1980 has been 0.37^{0}C. Actual temperature has fallen by 0.11^{0}C since 1998.

The graphs are worthless even without consideration of their flawed data sources.

This apocalyptic extreme is again derived, as had been the Mann hockey stick, from tree rings, boreholes and ocean cores. It is therefore necessary to test the validity of these more closely. Data sources and data for planetary latitudes are each examined.

PAGES2K data source

The 2021 hockey stick is derived from the work of the 'PAGES2K Consortium'. This is a group which is dedicated to reconstructing proxy based data to estimate temperature of the Earth over the last 2,000 years – hence **PA**st **G**lobal chang**ES** and the 2K. It has been producing circulars since 2010. It has many authors. Its compilations are the centrepiece of the IPCC *Summary for Policymakers*.

As explained proxy data comprise sources of temperature indications other than direct measurements. They are physical characteristics of the environment preserved to varying degrees. Proxies are natural indicators of climate variability and include such things as tree rings, ice cores, fossil pollen, plankton, ocean sediments, and corals. These allow reconstructions to be made of estimated temperature levels.

The validity of the proxies used by PAGES2K Consortium is so seriously compromised that they cannot be regarded as more than a further attempt to disguise known temperature variations of the last 2,021 years.

The principal flaws are described and scrupulously explained in analyses[134] prepared by Stephen MacIntyre of Climate Audit. It was he who first exposed the falsehood of the Mann hockey stick.

Proxy data flaws

Proxy data have extremes of inconsistency. That fact explained how Mann was able to justify his graph. Arkansas bristlecone pines showed what he needed to fabricate his graph but Californian pines did not.

The proxies which have been selected to support the new IPCC hockey stick graph consist of just 257 proxies. These have been selected from a dataset of 692 proxies. Those 692 proxies had in turn been selected from thousands of proxy series accumulated by many authors over the years.

Examination of random samples of proxy sets used in PAGES2K are reproduced in Climate Audit circular dated 2nd September 2021[135] . The batches of random proxy sets taken from the underlying series of proxies relied on show marked inconsistencies.

The following are reviews of sets of 11 random proxy samples of PAGES 2017. The series carried forward to PAGES2019 are in blue. The others do not appear.

---

[134] Climate Audit August and September 2021. PAGES2K: North American Tree Ring Proxies. Antarctic Proxies. Milankovitch Forcing and Tree Ring Proxies. IPCC AR6 [Assessment Report 2021] Hockey Stick. All of this section is taken from the brilliant analysis of Stephen MacIntyre to which references should be made.

[135] Climate Audit PAGES2K 0 – 30S.

Figure 30

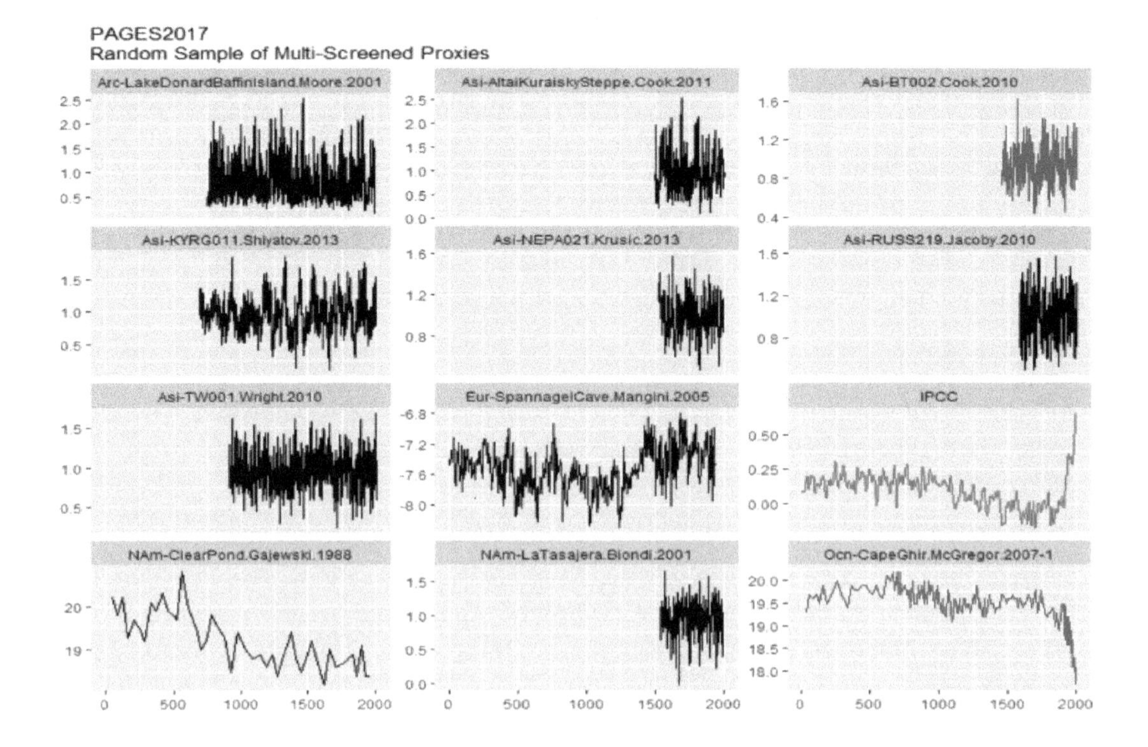

The downward trend of temperature revealed by the Cape Ghir (offshore Morocco) series is derived from alkenones (haptophyte algae) which, as has been explained in Section IV, are widely used to estimate ocean temperatures over millennia, being consistent and constant. The series coloured blue were carried into PAGES2019. The Cape Ghir data presented a quandary requiring explanation[136]. **However as MacIntyre shows the authors simply reversed the data by orienting the series according to its correlation with the targeted instrumental temperature.**

Tree ring data: strip bark

Tree rings have to be used with great caution. Leaf area, seasonality, solar radiation, canopy position, variation in evaporation, altitude, atmospheric salinity are local or regional variables.

The dependence of the selected PAGES2K proxies on tree ring data to support the global warming theory undermines the overall validity. The Report of Dr Gerald North for the US National Academy of Sciences as to the Mann hockey stick concluded that use of tree rings from bristlecone pines was so suspect that such proxy data *'should be avoided for temperature reconstructions'.*

But this direction has been disregarded.

Manipulation of tree ring data has been carried into the PAGES2K proxy series. The 'Mackenzie Delta' series from the Yukon, Canada show a 'super-stick' (2013). But ring widths did not correspond to the increase in temperature. Accordingly the framers of the proxy dataset concealed the decline in ring widths and temperature by excluding 'divergent portions' of the individual trees that had decreasing growth in recent years. MacIntyre includes a copy of the relevant entries in the proxy text and graphs of the impact of this distortion[137].

Paki 033

20% of the PAGES 2019 proxies are 50 Asian tree ring chronologies. Included is a series which do show an extreme closing hockey stick uptick (Asia_207). The data is cited to the site denoted by NOAA as paki033, a site located in mountainous northern Pakistan.

---

[136] Stephen MacIntyre Climate Audit 11th August 2021 03:14 for the source of all this information.
[137] 'The IPCC AR6 Hockey Stick' Climate Audit 11th August, 2021 – 3:14 PM.

Applying the measurement data for paki033.rwl at NOAA it was possible to calculate a site chronology using the most recent scientific expertise on chronology of tree ring dating (Dendrochronology).[138]

Figure 31

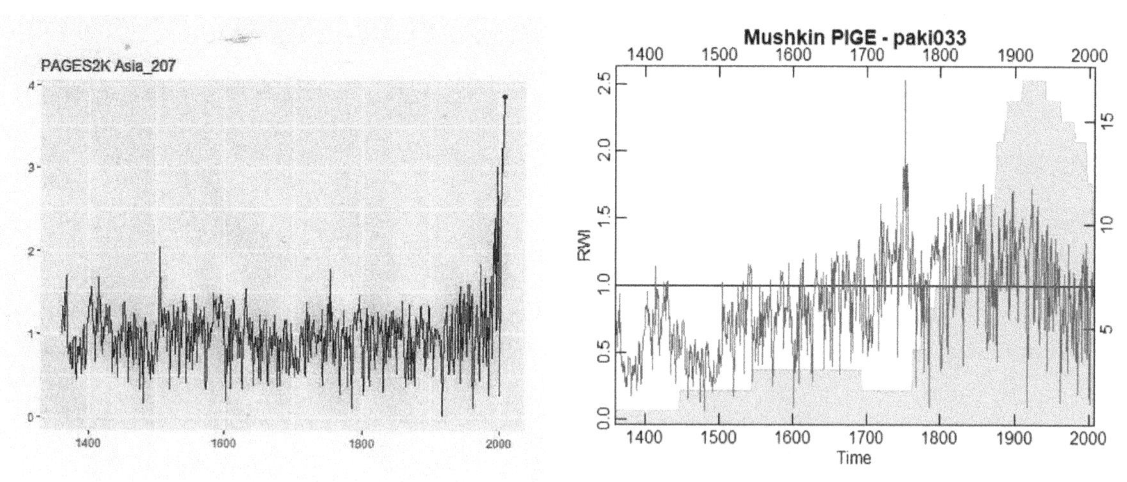

The hockey stick does not appear. **There is nothing in the actual ring width measurements that justifies the huge upspike in the archived PAGES2K Asia_207 chronology.**

PAGES2K continues to use most of the stripbark bristlecone pine data underpinning the Mann hockey stick. As MacIntyre observes *"the PAGES 2017 North American tree ring network has been severely screened ex post facto from a much larger candidate population"*. Over the years only 15% of the underlying population of North American tree ring network was selected. This was in order to impart the hockey stick blade. **Even so the composite of all data other than stripbark bristlecones had no noticeable hockey stick appearance**.

0⁰ – 30⁰ South latitude

Macintyre reproduces the proxies used for the southern tropical region (0⁰ – 30⁰S latitude).

The PAGES2K 0⁰ – 30⁰S proxy network has 46 proxies (there are just 8 proxies for 30⁰S - 60⁰– see below). However it has only **one** proxy from ocean core. For the period prior to 1500AD there are only two proxies with values prior to AD1500: the ocean core Makassar Strait and the ice core d18O series from Quelccaya, Peru. Indeed, there are only two other 0-30S proxies that begin prior to 1600 AD.[139] **There is no hockey stick trend as can be seen from the graphs of these proxies below.**

Figure 32

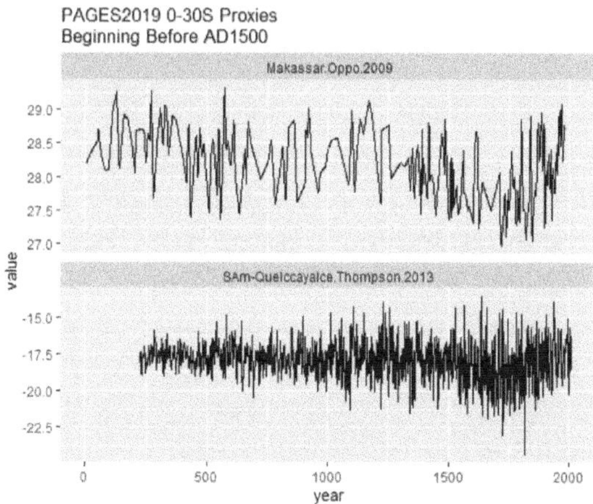

42 of the 46 proxy series are of the very short coral series. Half of them start after 1850 AD and no less than 30% after 1890 AD. One series (Clipperton Atoll, Wu [2014]) begins in AD1942. None of these

---

[138] A dendrochronology program library in R (dplR)." _Dendrochronologia_, *26*(2), 115-124. ISSN 1125-7865, doi: 10.1016/j.dendro.2008.01.002 (URL: https://doi.org/10.1016/j.dendro.2008.01.002). An Introduction to dplR Andy Bunn Mikko Korpela Processed with dplR 1.7.2 in R version 4.0.3 (2020-10-10) on 29th January 2021.

[139] Hendy (2002) Great Barrier Reef temperature reconstruction and an Indonesian tree ring series (INDO0005) that is non-descript in the underlying measurement (rwl) data at NOAA.

short series shed any light on whether the Medieval period was warmer than the modern period much less the Roman period.

PAGES2K only used one ocean core in the <u>0°-60°S</u> latband but omitting high resolution alkenone ocean series which are isotope related and reliable. Only four PAGES2019 series in the 0°-60°S band start prior to AD1100 and none of them have a hockey stick shape[140].

30° – 60° South latitude

The proxies used to cover the vast area of 30° - 60° latitude are reconstructed to show a hockey stick in PAGES 2019. This is shown below in Figure 32 (right hand graph).

But this is obtained by simply applying a coordinating multiple of 4 times to the previous chronological sequence. There is no comparable deviation in any of the underlying proxies.

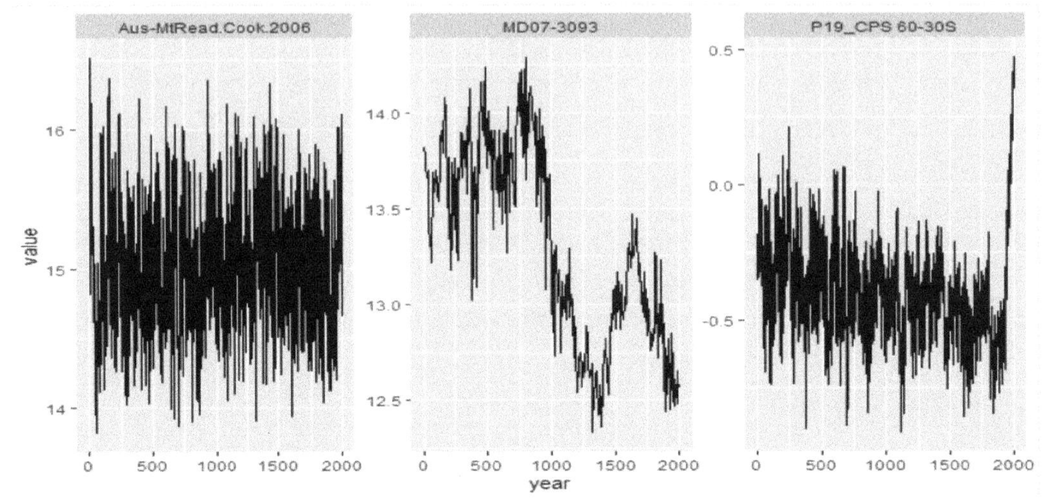

The above record shown on the left is the PAGES2K proxy for the longest period of the tree ring sample[141] with the PAGES2K reconstruction on the right. The central graph is the very first high resolution 30°S - 60° – ocean core alkenone ocean proxy[142] (2019). It has values dated from 372BC to 1992AD. It shows a 1°C to 1.5°C decrease from the first millennium up to and continuing in the 20th century.

But the astonishing fact is that all there is are just **7 (seven)** tree ring series. These doubtful proxies are all there is. The planet's surface in these latitudes is 96% ocean. . **Not a single proxy from the ocean has been used.**

That PAGES 2019 did not use any ocean core proxies fundamentally vitiates any conclusion drawn from tree ring and the very short coral series. As has been explained above (see Section IV 'Roman Warming Period'), there are physical formulas for estimating sea surface temperature from alkenone or Mg/Ca measurements. As MacIntyre explains[143] any conversion of tree ring widths to temperature in deg C is simply the result of ad hoc statistical fitting, not by application of a universal formula. As he states *'alkenone values have been measured all over the modern ocean and nicely fit known ocean temperatures. In addition, alkenone values for ocean cores going back to deeper time (even to the Miocene[Epoch[) give a consistent and reproducible narrative'.*

---

[140] Stephen MacIntyre Climate Audit September 2 2021.

[141] This is the only 30-60S proxy series in PAGES 2019 that reaches back into the first millennium – the Mount Read, Tasmania tree ring.

[142] Collins JA et al: *Centennial-scale SE Pacific sea surface temperature variability over the past 2,300 years,* 2019.

[143] Stephen MacIntyre Climate Audit September 2 2021.

0° – 30° North latitude[144]

The IPCC reconstruction for this latitude band looks almost exactly the same as reconstructions for the 0°-30°S and 30°-60°S (see above) **However, none of the <u>actual</u> proxies in this band look remotely like the band reconstruction.** In prior PAGES2K compilations (2013 and 2017) there were significant long term proxy values. These have been virtually eliminated and very short term coral proxies included (71%). All but 3 of the coral series begin after 1775 with many in the late 19th or even 20th centuries. As a proportion of the network, corals went from 1.75%, 1 of 57 proxies in the PAGES13 network, to 71% of the PAGES2019 network (29 of 41).

The primary purpose of PAGES 2000 year proxy reconstructions of temperature is to compare modern temperature to estimates of medieval and first millennium temperatures. There are just 41 proxies in the 0-30N network, but only three proxies with values before 1200 AD and only **one** proxy with values prior to 925 AD (see graph below).

The single long proxy with values through the first millennium is a temperature reconstruction from Mg/Ca values from an ocean core taken from the sea bed located offshore from North Africa. It is intermittent and erratic over two millenia with minor recovery in 20th century.

Figure 34

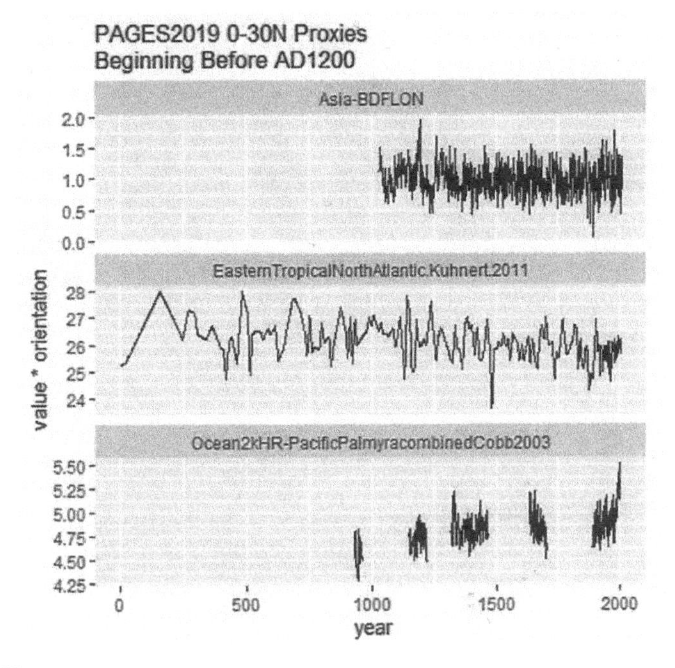

Tree rings 0° – 30° N

PAGES2013 had 54 tree ring chronologies in the 0° – 30°N band. This network was virtually unchanged in PAGES2017. However in PAGES2019, the network of Asian tree ring series chronologies is cut to just eight (see below). **None of these chronologies are similar to the IPCC reconstruction.**

Two of them (CENTIB, MAXSIC) have very late upspikes – a phenomenon examined by MacIntyre recently in connection with another Asian tree ring chronology (see above) for which underlying measurement data had become available. It was impossible to replicate the upspike with usual chronology techniques and there was no apparent basis for the upspike existing in the underlying measurement data. In response to a recent inquiry by MacIntyre, the PAGES2019 authors were unable to identify how the chronology was calculated and refused to find out.

[144] See for all this Stephen MacIntyre exhaustive analysis Climate Audit Sep 15, 2021 – 12:01 PM.

Figure 35

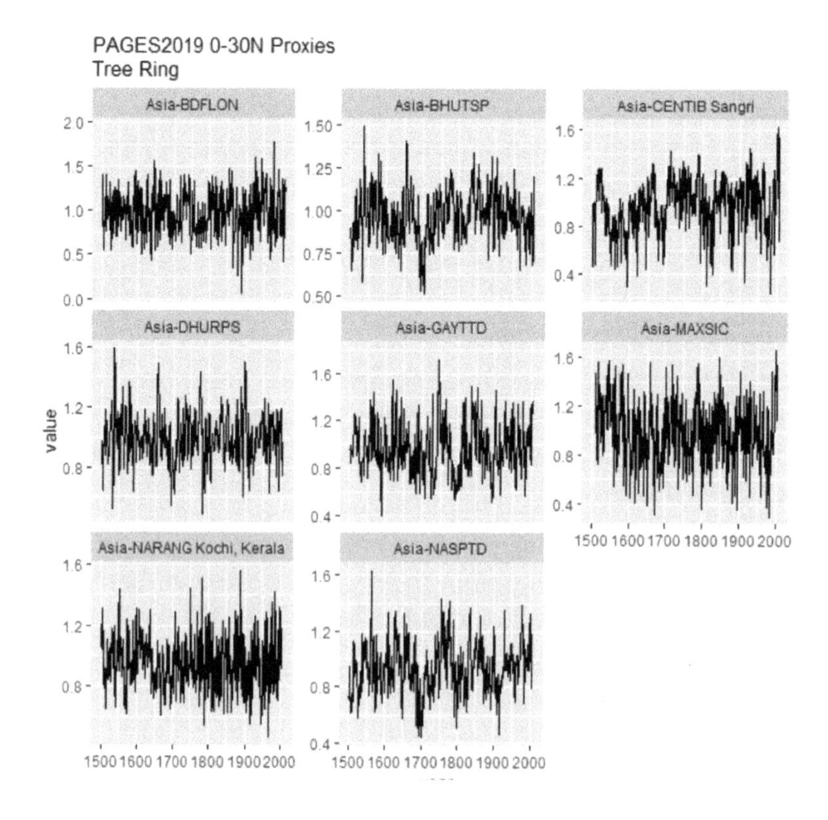

Ocean Cores

The number of ocean cores in the 0^{0} – 30^{0}N band was cut even more severely -: from 21 series in PAGES2017 to just 3 in PAGES2019. Only **one** of these is a proxy with long values. The graph below shows eight PAGES2017 ocean proxies with values prior to AD1000. Only the Kuhnert 2011 in red below was retained in PAGES2019. **Except for the Dry Tortugas series (Lund et al 2006) the others show declining or stable values through the last millennium**.

Figure 36

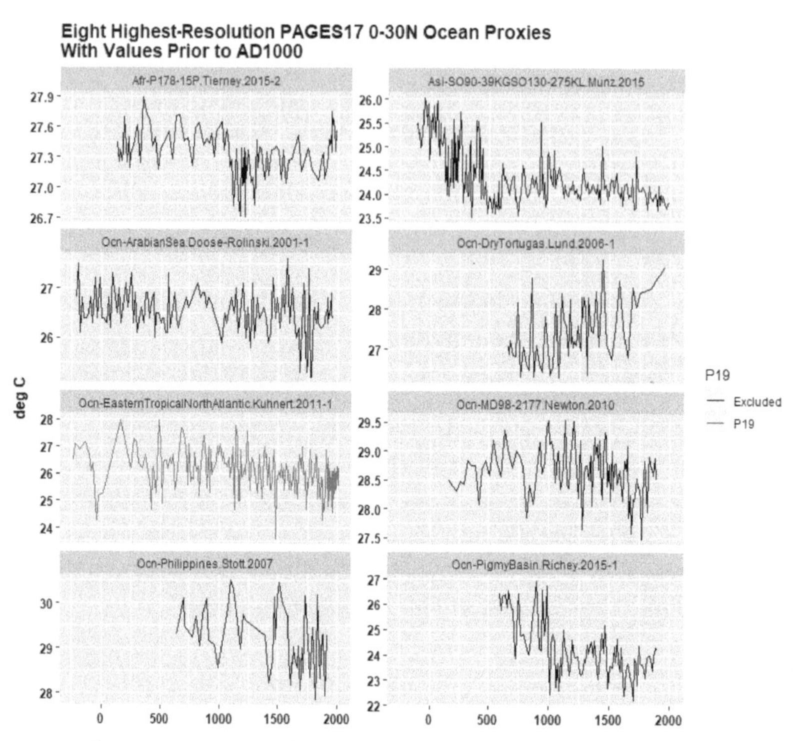

The PAGES2019 network contains two other shorter ocean series – neither of which reach prior to AD 1200. The removal of 18 of 21 ocean cores removed most of PAGES2017 "long" proxies.

d18O Proxies $0^0 - 30^0$ N[145]

The following graphs (**Figure 37**)are derived from oxygen isotope data.

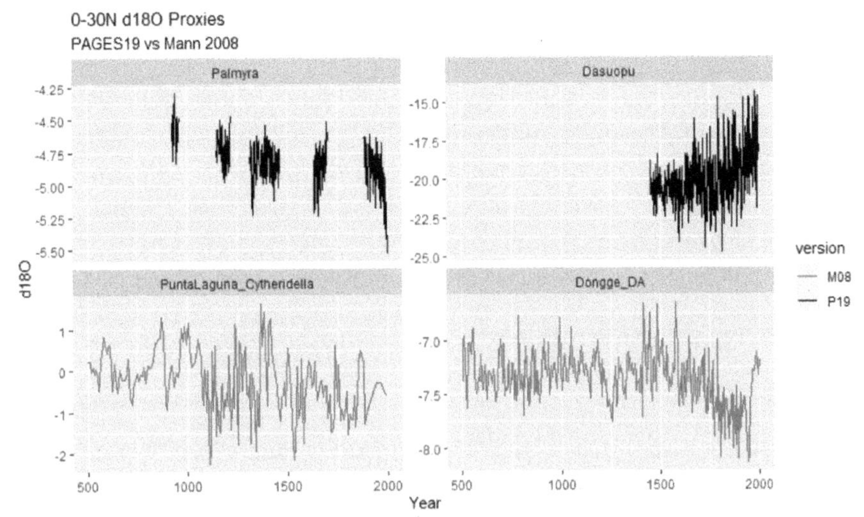

PAGES 2019 contained 2 "non-short" 0^0-30^0N d18O proxies (in addition to myriad very short coral series): Palymra coral (6N, 160W) in the heavy ITCZ rainfall[146] band; and a truncated version of the Dasuopu ice core. **But it eliminates the long term proxies of earlier PAGES including the 2008 Mann et al. These show a contrary trend.**

In the above graphs by way of comparison, Macintyre shows two long d18O series – a speleothem (Dongge, China) and a lake sediment series (Yucatan, Mexico) from the longer series which had been included in 2008 Mann et al. **Both series show declining values in the last two millennia.**

Conclusion[147] *as to $0^0 - 30^0$N proxies*

A temperature proxy must be linearly related to temperature. Accordingly proxies in a network of actual temperature proxies should have a reasonable consistency which should also appear like the reconstruction. **The IPCC reconstruction does not reflect the PAGES2019 network for $0^0 - 30^0$N.**

Moreover, the proxies covering the medieval period and earlier are absurdly sparse. Such proxy series have been widely available in the past 15 years yet PAGES 2019 contains only **one** proxy for prior to AD925. Not only this, it **actually reduced** the representation of longer (ocean core, speleothem, lake sediment) proxies from Mann et al 2008 and PAGES2017, while dramatically increasing the proportional representation of very short coral proxies.

Let Stephen MacIntyre conclude:-

*"Finally, the network is wildly inhomogeneous over time. In the past two centuries, it is dominated by trending coral proxies, with only a few nondescript or declining long proxies. Any form of regression (or like multivariate method) of trending temperatures against a large network in the instrumental period will yield an almost perfectly fitting reconstruction in the calibration period if the network is large enough. But when the network is limited to the few long proxies (and especially the singleton proxy extending to the first century), the fit of the regression (or multivariate method) will be very poor and **the predictive value of any reconstruction negligible**."*

[145] Ratio of stable isotopes oxygen-18 (^{18}O) and oxygen-16 (^{16}O). In paleosciences, ^{18}O:^{16}O data from corals, foraminifera and ice cores are used as a proxy for temperature.
[146] Intertropical Convergence Zone.
[147] See for all this Stephen MacIntyre exhaustive analysis Climate Audit 15th September, 2021 – 12:01 PM.

Conclusion re Proxy basis of 2021 graphs

The 2021 IPCC graphs are a perpetuation of the distortions or truncations of proxy derived data that were first revealed by the 1998 Mann hockey stick. They seek to eliminate the 'inconvenient' variations in the temperature of the planet that have occurred in the last 2,020 years. They posit a vertiginous leap over the past three decades.

Such has been the disgraceful means by which the IPCC has sought to re-constitute climate history. It has been well said that *"He who controls the past controls the future. He who controls the present controls the past"*. [148] Wherever the 2021 IPCC 'report' diverges from original data or from satellite and balloon sonde measurements it should be utterly disregarded . It is not science. It is information of a biased and misleading nature used to promote a political cause – it is propaganda.

Misrepresentation as to peak temperatures

Hottest for 125,000 years

The IPCC graph and related text make claims that current temperature of the Earth is higher than it has been for 125,000 years. This was in the warm period of 14,000 years of 'interglaciation' from 130,000 to 116,000 years ago. Of those 125,000 years 104,000 years fell within the full glaciation (Ice Age) ending about 14,000 years ago. The IPCC has sought to exaggerate, by obvious implication, the contextual significance of increase in the Earth's temperature since that time. Moreover it asserts that today we are in the warmest time for 6,500 years a period following the Holocene Warming (8,100 – 6,500 years ago). Even this assertion is flawed.

Figure 38

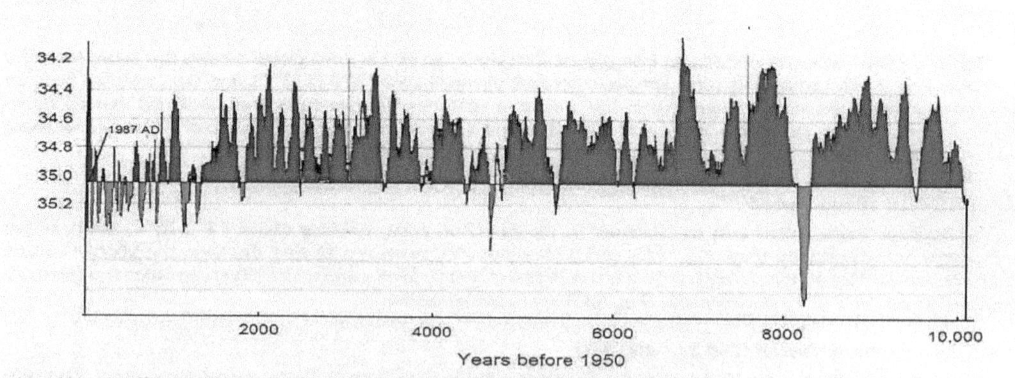

FIGURE 8.10 $\delta^{18}O$ from the GISP2 ice core for the past 10,000 years. Red areas represent temperatures warmer than those in 1987 (top of the core); blue areas were cooler. Almost all of the past 10,000 years were warmer than the past 1500 years. *Plotted from data by Grootes, P.M., Stuiver, M., 1997. Oxygen 18/16 variability in Greenland snow and ice with 10^{-3}- to 10^{5}-year time resolution. Journal of Geophysical Research 102, 26455–26470.*

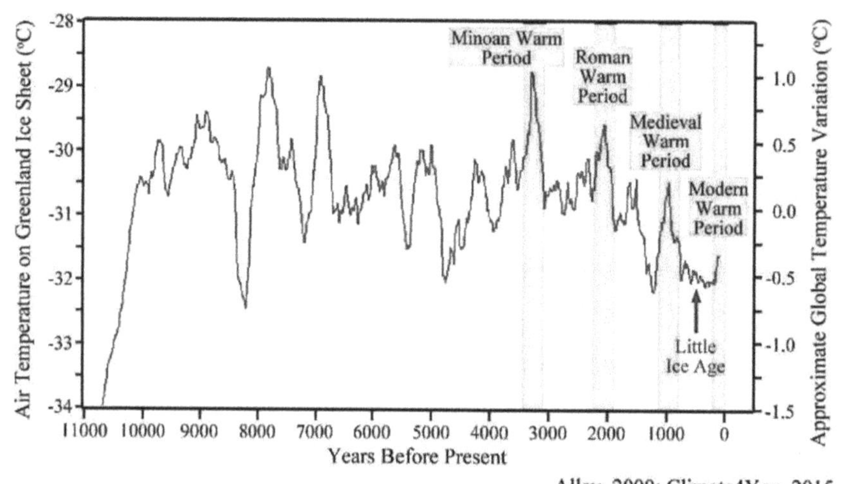

Alley, 2000; Climate4You, 2015

---

[148] George Orwell *Nineteen Eighty Four.*

The first graph in Figure 38 is is taken from Professor Easterbrook's work[149]. It shows an oxygen isotope record which begins in 1987. Temperatures that were higher than in 1987 are shown in red. The stand out feature of the graph is that temperatures for the entire record have for the most part been higher than today save for minute variations and the 400 year period of 8,900 – 8,500 years ago [150]. It is the case that global temperature levels were between 1 – 3°C higher for almost the entire period of 10,000 years.

The second graph in Figure 38 is also taken from the Greenland isotope record[151] and shows the relative highest temperature peaks of the Holocene Optimum 8,000 – 6,500 years ago, the more recent Minoan Warming, Roman Warming and Medieval Warmings in relation to the Modern Warm period to 2015.

It is important to recall that the satellite record (Figure 11) shows that over the 34 years since 1987 temperature has risen by just 0.4°C and that the total rise in average temperature since 1850 to 2021 was just 1.1°C and arguably less[152].

Unprecedented in more than 2,000 years

This statement is consistent only with the unwarranted and misleading eliminations of both the Roman Warming period 250BC to 450AD and also the Medieval Warming period 900AD to 1300AD.

It is contradicted by the evidence. We have already examined the compelling evidence of high temperatures in these important periods (Figures 16 and 17).

However it is instructive to consider what actually happened during those times.

The Roman Warming period lasted from 250BC to 450AD. Citrus trees and grapes were grown close to Hadrian's Wall. By 300AD global climate was far warmer than at present[153]. Populations increased with higher agricultural yields and prosperity. Tree lines reached much higher altitudes. This warming and its effects were a global phenomenon as is clear from scientific studies[154]. The warming coincided with intense solar activity.

The Medieval Warming of approximately 400 years saw a resurgence of prosperity, population growth and plenty. Surplus food in Europe led to great increases in population. The Oxford and Cambridge colleges have their origins in this time. The foundation stones of almost all the great Medieval cathedrals were laid in this period. Cattle, sheep, and barley were grown in Greenland. No less than 650 farms of those times have been excavated in Greenland. Glaciers retreated and Arctic ice receded. Temperature was higher than at any time until the spike of 1998 (Figures 17 and 18).

The IPCC reports stipulate that global warming increases should be limited to 2°C from pre-industrial levels. Yet for most of the Holocene they were higher than this and each has resulted in periods of prosperity, expansion and well being for humanity[155].

---

[149] Professor Easterbrook Op cit p 143.

[150] This was caused it is believed by collapse of vast meltwater dams See Plimer Op Cit p 48 and authorities cited.

[151] Air temperature at the summit of the Greenland Ice Sheet, reconstructed by Alley (2000) from GISP2 ice core data. Climate 4you.com.

[152] See "170 Years of Earth Surface Temperature Data Show No Evidence of Significant Warming" Author: Thomas K. Bjorklund, University of Houston, Dept. of Earth and Atmospheric Sciences, Science & Research.

[153] See e.g 100 – 400 alpine record AD Röthlisberger, F. (1986), *10,000 Jahre Gletschergeschichte der Erde*, Sauerländer, ISBN 978-3794127979. Bianchi GG, McCave IN; McCave (February 1999), "*Holocene periodicity in North Atlantic climate and deep-ocean flow south of Iceland*", Nature, 397.

[154] See authorities cited in Plimer Op cit pages 59-60.

[155] *"For the first time, we can state the Roman period was the warmest period of time of the last 2,000 years, and these conditions lasted for 500 years"* Professor Isabel Cacho at the Department of Earth and Ocean Dynamics, University of Barcelona. July 2020.

Conclusion

It is no less than the truth to say that the IPCC has wilfully misrepresented the evidence of the original accepted data records. On these false statements and graphs rest decisions by our Government which will require expenditure of at least £3 trillion, most of it in the next few years.

SUMMARY

The 2021 *Summary for Policymakers* knowingly misrepresents the global temperature record over modern times and the geological past. For almost all of the last 10,000 years temperature was higher than today. The IPCC graphs, with intent to deceive, knowingly exclude pronounced historical warming and cooling variations inconsistent with its dogma.

Had directors of a company issued a prospectus with these misstatements inviting the public to subscribe funds for such a colossal venture as replacing fossil fuels in their entirety by 2050 they would have been liable to prosecution for offences under the Financial Services legislation.

APPENDIX

10 SPECTRES OF THE CONSENSUS

What happened last time?

We live in a geological period known as an interglacial. It began about 14,000 years ago. Such periods last 10 – 15,000 years. We are near the end of this interglacial. The last complete interglacial was between 130,000 and 116,000 years ago. During that time temperature was up to 6⁰C warmer at the poles and 2⁰C warmer at the equator[156]. Sea level was 4 – 6 metres higher than it is today. Glaciers retreated and tree lines rose in altitude reaching up into mountains. Tundra was replaced by trees. Sea temperatures rose. Coral reefs rose and thrived between 128,000 and 121,000 year ago at the height of that era's temperature – well over today's level – in areas now far too cool for coral[157]. During that interglacial *homo sapiens* evolved in East Africa though a primitive version had emerged about 300,000 years ago. Polar bears did not become extinct. Polar ice sheets did not completely melt.

The 12,000 warm period was not caused by CO_2 concentrations but by the Earth's eccentric orbital cycle intensifying solar radiance.

We constantly hear of bush fires, forest fires, floods, hurricane damage, and other natural events. These are disasters for those affected. But they are no more indicative of a trend of global warming than ripples on the sea are of the tides much less the ocean currents[158].

Belief that humanity is responsible for these reveals a blind acceptance of the 'consensus' dogma. The truth is that these events are the recurrence of tragedies that have been the lot of mankind for millennia. That some may have graver consequences is due largely to the explosion of the population of the planet rendering so many more lives and so much more property at risk.

In this Appendix is a summary of the principal adverse events that are attributed, mistakenly, to warming of the planet caused by the additional CO_2 emitted by humans.

The most fundamental and simple repudiation of these stories is that the temperature of the Earth is almost entirely explained by solar activity and its emergence from the Little Ice Age. Further increases in CO_2 will have negligible effect on temperature (see Section II).

Coral reefs are disappearing

Coral reefs are expanding[159] and contracting as they have for millions of years. The current scare is that the oceans are being made acid as a result of the minute increase (5% of 0.041%) in atmospheric CO_2 caused by human emissions of fossil fuels. Ocean acidity is measured as pH. It is now between pH 7.9 – 8.2. Neutral is pH 7. Thus the oceans are alkaline. To acidify oceans at a level of pH6 ten times more acid is needed than for pH7. To acidify seawater from 8.2pH to pH6 vast amounts of acid are necessary.

It is alleged that acidification results in loss of calcium carbonate used to create skeletons including corals. However the geological record shows that shells do not dissolve.

When CO_2 is dissolved in seawater it is converted to bicarbonate by reactions with dissolved carbonate and borate in water and with calcium carbonate sediment on the ocean floor. Furthermore, the oceans

---

[156] See the source of this text taken from Professor Plimer Op Cit 2017 at pp 36, 37 and detailed authorities cited.
[157] Stirling C.H. et al 1997 *Timing and Duration of the Last Interglacial: Evidence for a restricted interval of widespread coral reef growth*.
[158] See for example the short article in the *Daily Telegraph* 11th August 2021 by Amber Rudd which simply assumes global catastrophic warming because of these local perils and tragedies. She displays no understanding whatsoever of the drivers of climate change or the properties of CO2…."*we are seeing the impacts of climate change unfold around us with terrifying ferocity. Wildfires in California, Greece and Italy. Floods in China and Europe, and in our own towns and villages*".
[159] Dr Paul Kench Coastal Geomorphologist University of Auckland reported on Australian ABS the expansion of coral atolls in the Marshall Islands, Kiribati and Maldives archipelago by 8%: wattsupwiththat.com 10th January 2021.

are saturated with calcium carbonate to a depth of 4.8km. Thus any further CO_2 would precipitate calcium carbonate. At the level of the ocean floor the reaction of the extensive sheets of volcanic basalt and seawater renders seawater more alkaline by removing CO_2 form carbonates. In addition micro-organisms that consume CO_2 increase alkalinity.

There have been vast changes in CO_2 density since the Cambrian era which heralded the arrival of multicellular predatory creatures. Yet the oceans have not become acid. Fossilised shells, algal reefs, and coral reefs reveal that alkalinity was maintained at far higher temperatures.

Pollution and run off from land has had a considerable impact on coral reefs. In areas of the oceans which have no such impacts there are no signs of decline due to alleged acidification[160]

Corals recover rapidly from bleaching[161]. The 2016 bleaching was caused almost certainly by the sharp rise in global temperature caused by the El Niño event 2015/6. Recovery has started and is expected within 5 years[162].

Polar bears

The last interglacial 130,000 – 116,000 years ago was much hotter than today. The Holocene Maximum of 6,500 years ago was higher than today with long periods of ice free summers. Almost all of the past 10,000 years have seen warmer conditions than today (Figure34). The greatest threat to polar bears has been vile human hunting. Since this was controlled the population has increased from about 10,000 to 26,500[163]. The bears do not need ice to survive. But the hunting goes on.

Arctic ice

Because of the absence of any land mass in the Arctic Ocean, most of area lacks glaciers, which require a land mass. Thus, most of the Arctic contains only floating sea ice. Greenland, Iceland, northern Alaska, and northern Siberia contain the only glaciers in the general Arctic region.

Arctic warming and the melting of the arctic ice happen predictably on multidecadal scales with a period of around 60 years and are entirely natural. Warming results in part from the reduction of arctic ice extent because of flows of warm water from the Pacific through the Bering Straits and from the Atlantic. The warmer water thins the ice from beneath, slows the refreezing and limits degrees the depth and extent of the ice. Arctic ice floats on water and the melting has no effect on sea level.

However, the driver for the multidecadal ocean cycles is almost certainly the solar cycles. As noted in Dr Soon[164] showed how the Arctic temperatures (Polyakov) correlated extremely well with the total solar irradiance but not with CO_2. This is revealed in the following graphs.

---

[160] Woods Hole Oceanographic Institution 27th August 2020.

[161] The process of bleaching occurs when coral expel algae on which it depends. This is due to a variety of factors including even slight temperature changes as little as 1.12C, extremely low tides, pollution, or too much sunlight.

[162] See article *"Marvellous Resilient Coral"* 5 October 2019 wattsupwiththat.com . See also the *Phoenix effect* of coral recovery described by Dave Krupp a coral researcher in Hawaii who works with *Fungia scutaria*. Bleaching does not mean death. 2016 bleaching of the coral in the Coral Sea Marine Park did not result in any significant loss of coral cover.

[163] Polar Bear Research Group.

[164] Dr W Soon astrophysicist at the Solar, Stellar and Planetary Sciences Division of the Harvard-Smithsonian Center for Astrophysics. Founder of Center for Environmental Research and Earth Sciences (CERES-science.com).

Figure 39

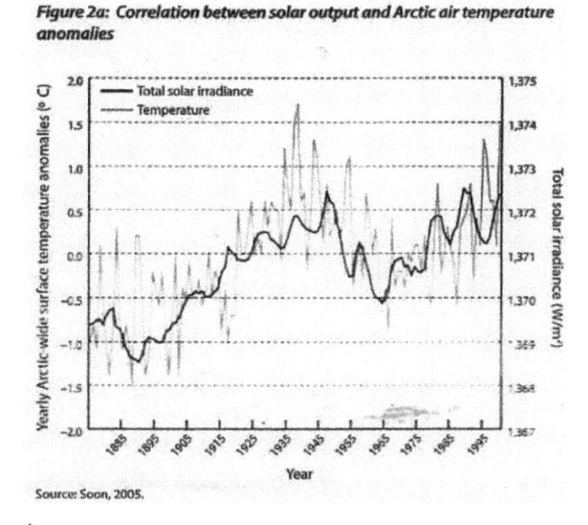

Figure 2b: Much weaker correlation between atmospheric CO_2 and Arctic air temperature anomalies

Carbon dioxide (CO_2)

Temperature

Yearly Arctic-wide surface temperature anomalies (° C)

Atmospheric carbon dioxide (ppm)

Year

Source: Soon, 2005.

www.fraserinstitute.org • Fraser Institute

Greenland

Greenland's climate changes relate to ocean cycles on large scales. Ice core data reveal 1,000-year cycles, slowly declining as we reach the end of our present interglacial period. The Medieval Warm Period saw ice retreat in Greenland enough to entice the Vikings to settle Greenland and grow crops including grapes for wine. They abandoned it with the onset of the Little Ice Age.

Figure 40

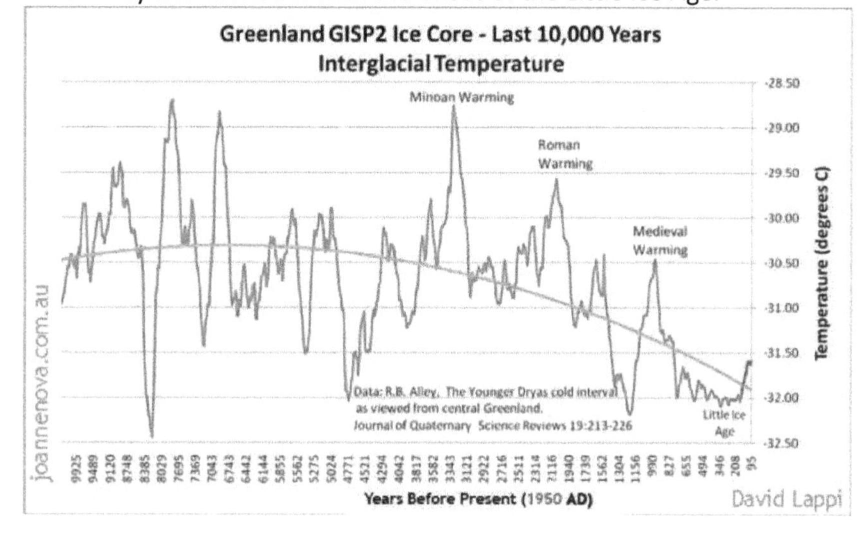

Antarctic Ice

The figure below shows temperature changes during the past 30 years, compared to temperatures recorded between 1950 and 1980. Most of Antarctica is cooling, with warming occurring over just a small portion of the continent that juts out into the Southern Ocean (the Antarctic peninsula). That region is volcanic with both surface and sea bottom vents. It is also subject to periodic warming from winds that blow downslope and locally warm by compression.

Figure 41

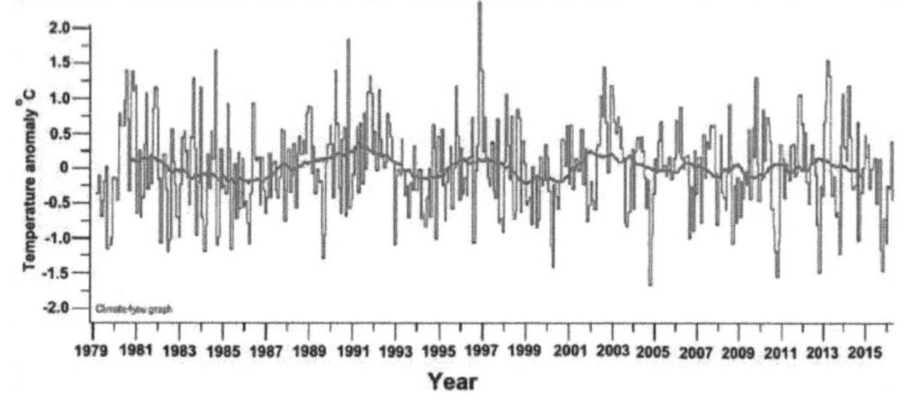

Sea levels

Hotter sea water expands in volume. Melting land based ice adds water and this includes glacier melt and the Antarctic peninsula discussed previously.

Following the end of the last glaciation (12,000 years ago) sea levels rose by 130 metres. But this was almost all within the first part of the interglacial. It reached present levels 7,700 years ago. It remained 1.5 metres higher than today until approximately 3,000 years ago when it fell. Sea rises and falls of 2 – 4 metres over several decades have been common in the last 6,000 years.

There are very many variables that affect sea levels.

Global rises in sea levels due to heating of the ocean surface take a long time to have an effect following the rise in surface temperature. Sea levels in the Roman Warming Period were well over 1 metre lower than today. It has also been established that coral atolls rise and fall in response to the rise and fall of the volcanic substructure.[165]

Warmer sea water means greater evaporation. This moisture is deposited in cold regions. This will affect the rate at which sea level rises and may lead to a fall in sea level. It has been noted that a l°C rise in the atmospheric temperature of Antarctica would decrease sea level by 0.2 to 0.7mm per year due to the slow overturning and currents of the deep ocean and increased precipitation adding to the ice cap.[166]

There are many factors bearing upon sea level changes to be able to have any certainty as to the causes and their relative significance. So much is of local and regional impact not reflected globally. But what is certain is that the gradual rise in sea levels is entirely consistent with the emergence of the Earth from the Little Ice Age. The most that can be said is that since 1880 sea levels have risen at the rate of between 13mm and 20mm every 10 years. This is entirely consistent with the gradual rise with our leaving the Little Ice Age.

Moreover there is a close correlation between solar variability and the variability of sea levels and levels of great lakes.

Figure 42

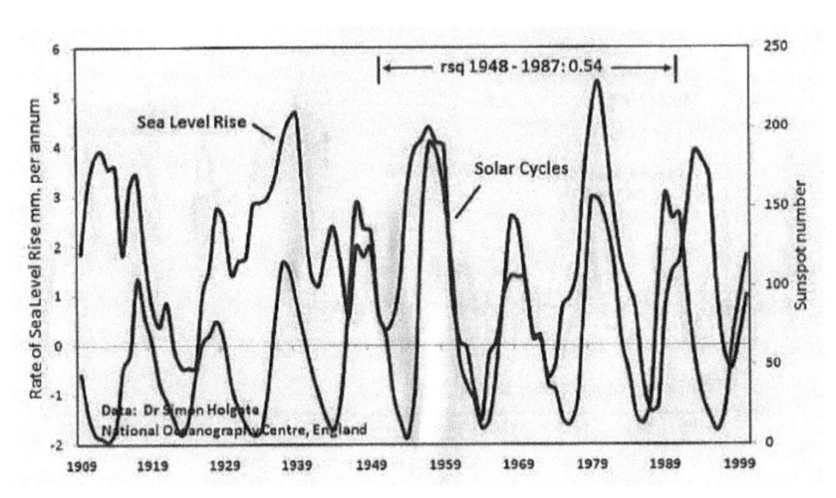

Human contribution to this is essentially speculative given the time lag between heating of the oceans and temperature rise. However even assuming that (which is not the fact) there is no such time lag the human contribution has been estimated at 7mm every 10 years[167].

---

[165] Professor Plimer Heaven and Earth Op cit 2009 pp 320,321 and authorities cited.

[166] Reeh N 1999 Geografiska Annaler 81A 735 – 742 cited in Professor Plimer 2009 Op cit. page 313.

[167] Professor Roy Spencer University of Alabama and Church, J. A. and N.J. White (2011), "Sea-level rise from the late 19th to the early 21st Century", Surveys in Geophysics, 32(4-5), 585-602, doi:10.1007/s10712-011-9119-1.

Drought. Desertification

The incidence of drought in the USA since 1890 is set out in the following graph. It needs to be borne in mind that reliable evidence of drought is readily available from the statistical record and the continental USA has many and varied climate zones.

Figure 43

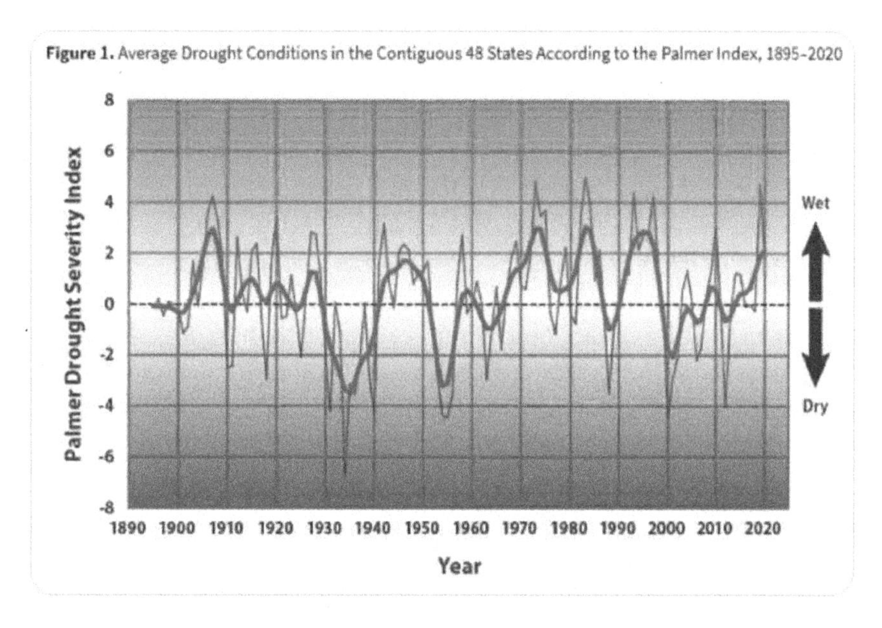

Note the correlation of severe drought peaks with the sharp rise in temperature of the 1930s, the decline in the cooling period after 1945, the rise in 1976, the El Niño spikes of 1983, 1998, and of 2015/6. There is no correlation whatsoever with CO_2 emissions. Nor is there evidence of intensification either in scale or degree.

Even the US National Climate Assessment[168] concluded that there had been no increase in drought in the USA over the historical record[169].

The IPCC falsely asserts that 'climate change' (meaning warming) contributes to desertification[170]. Desertification is a consequence of cooling not warming. Dry cold air and the reduction in CO_2 and rainfall bring it about. Rainfall and increased CO_2 bring about greening (see Figure 2).

The greening of the lower Sahara has been documented (see section on Lake Chad below). Deserts are also on the retreat in China[171].

---

[168] The NCA and its assessment reports mimic the IPCC and its reports. It assumes that the warming of I.1°C since 1850 is mainly attributable to human emissions of greenhouse gasses. See for example its concealment of the true data grounded incidence of hurricane S V Koonin "Unsettled" Op cit pages 116,118 which is illustrated in the graph of the North Atlantic Power Dissipation Index below.

[169] https://nca2018.globalchange.gov/chapter/2/...." *the Dust Bowl of the 1930s remains the benchmark drought and extreme heat event in the historical record, and though by some measures drought has decreased over much of the continental United States in association with long-term increases in precipitation (e.g., see McCabe et al. 2017), there is as yet no detectable change in long-term U.S. drought statistics".*

[170] 2021 Summary for Policymakers "*Climate change, including increases in frequency and intensity of extremes, has adversely impacted food security and terrestrial ecosystems as well as contributed to desertification and land degradation in many regions (high confidence)*". {2.2, 3.2, 4.2, 4.3, 4.4, 5.1, 5.2, Executive Summary Chapter 7, 7.2}.

[171] See Plimer Op Cit 2009 pp204 – 207 and authorities cited.

Lake Chad

Reference is often made to the fate of Lake Chad once the 6th largest lake in the world. This has been due to entirely factors other than any supposed climate change by reason of of human emissions[172].

It has been due principally to the vast drawdown of water for agriculture.

As has been explained (Section IV) deserts are formed at times of global cooling and falling precipitation of rainfall[173] taken together with less photosynthesis and thus less CO_2. There is now evidence to support the claim that the southern border of the Sahara has been retreating since the early 1980s, making farming viable again in what were some of the most arid parts of Africa[174]. There has been a spectacular regeneration of vegetation in northern Burkina Faso, which was devastated by drought and advancing deserts 20 years ago. Rainfall and CO_2 increase are the causes of this beneficial transformation.

Hurricanes

The term hurricane is also used to describe cyclones or typhoons. They arise from tropical depression over the oceans close to the equator. 60% are in the Pacific, 30% in the Indian ocean. The 10% in the North Atlantic and specifically affecting Louisiana are depicted on the following graphs.

Figure 44

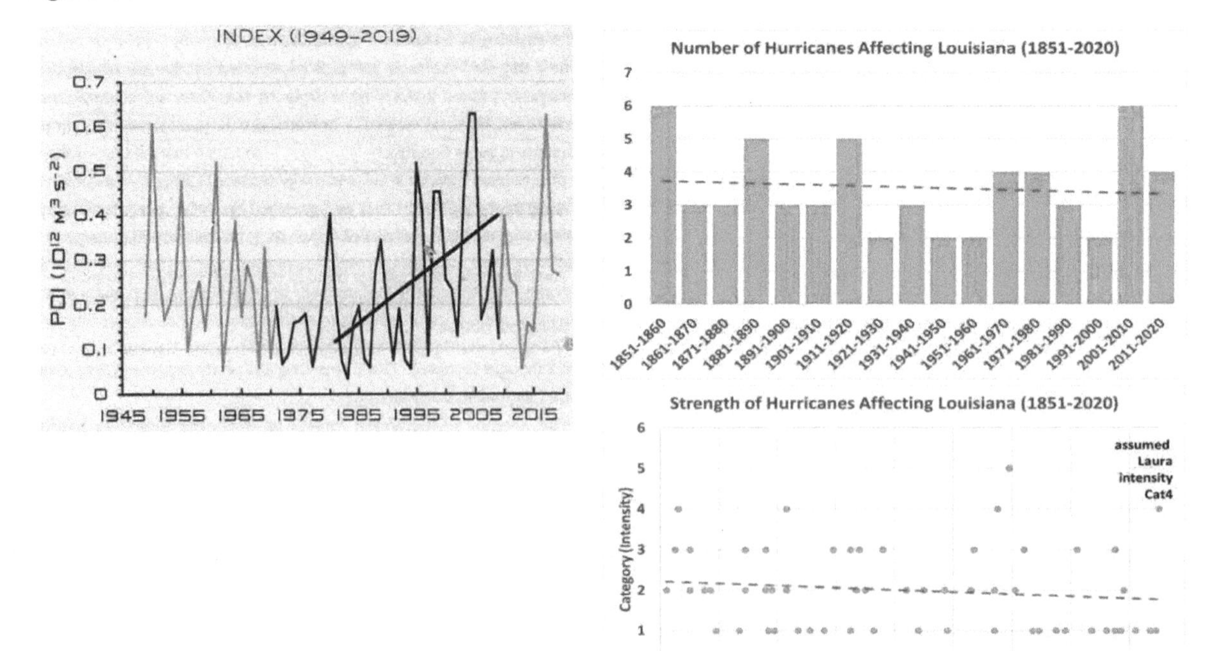

There is good observational evidence as to the incidence and strength of hurricanes since 1950 when atmospheric CO_2 concentrations increased sharply. The most severe hurricanes are rated 5 on the Saffir-Simpson scale with least severe at 1. However a number of factors other than surface water temperature govern variations in the Power Dissipation Index[175]. These include wind shear[176] and dust.

---

172 *"On the causes of the shrinking of Lake Chad"* H Gao et al 2011 Environ. Res. Lett. 6 034021 Failure of the lake to remerge with renewed rainfall in the 1990s following the drought years of the 1970s and 1980s is a consequence of irrigation withdrawals.

173 See Plimer Op Cit 2009 pp204 – 207 and authorities cited.

174 Phillip Mueller *"The Sahel is Greening"* The Global Warming Policy Foundation Briefing paper no 2.

175 A scientific measurement of storms weighted according to the cube of wind velocity.

176 Variation of wind speed or wind direction according to altitude.

It is clear from the data records that the number and strength of hurricanes affecting one of the most affected parts of the USA – Louisiana (right hand graph) – have shown no increase since 1851[177].These regional trends are reflected in the graph of the North Atlantic Power Dissipation Index set out above (left hand graph).

Extinctions

The so-called 'Extinction Rebellion' is a crude and violent form of extreme environmentalism the aim of which is a socialistic reversal of economic growth and overturning of its energy base.

It has nothing to do with any crisis of species extinction existing or probable. Oxygen caused the first major extinction 2,300 million years ago with the explosion of bacterial life deprived prokaryotic organisms of CO_2. Since then extinctions of varying severity have afflicted the biosphere throughout Earth's history. Supervolcanoes emit highly poisonous hydrogen sulphide oxidising in the atmosphere to sulphuric acid collapsing the food chain and dissolving the shells of marine life. Sulphur dioxide has similar effects. Basalt supervolcanoes have eliminated much of terrestrial life many times in the geological past.

Hunting by humans has over the past 40,000 years had a devastating effect and human colonisation has itself caused extinctions in particular by deforestation. However it is global cooling that leads to massive and repeated biological stress. Species diversity falls, rainfall diminishes, and there is a loss of shallow water habitats.

Warming is not associated with extinctions. Nor is global CO_2 density. There were no mass extinctions in the Holocene Maximum 6,500 years ago or the Minoan Warming period 3,500 years ago or the Roman Warming Period or the Medieval Warming Period. In those periods temperature was between $1^{0}C - 5^{0}C$ higher than today and for thousands of years about $3^{0}C$ hotter. Geology shows that in all previous epochs in times of global warming there is expansion of life. Animals and plants are limited by latitude and altitude in their range[178]. With warming further habitats become available. Moreover if CO_2 is doubled then at $10^{0}C$ plant growth is unaffected. However it is doubled at $38^{0}C$[179]. There is no basis on which geological history or the past few thousand years can sustain predictions of extinctions due to climate change much less that alleged to be due to human CO_2 emissions.

Heat deaths

Of all the weather extremes blamed on the global warming theory heat waves permit the most rapid and effective adaptation[180]. Moreover studies of heat related deaths do not at all reflect or suggest a causal connection with global warming induced by human activity.

Figure 45

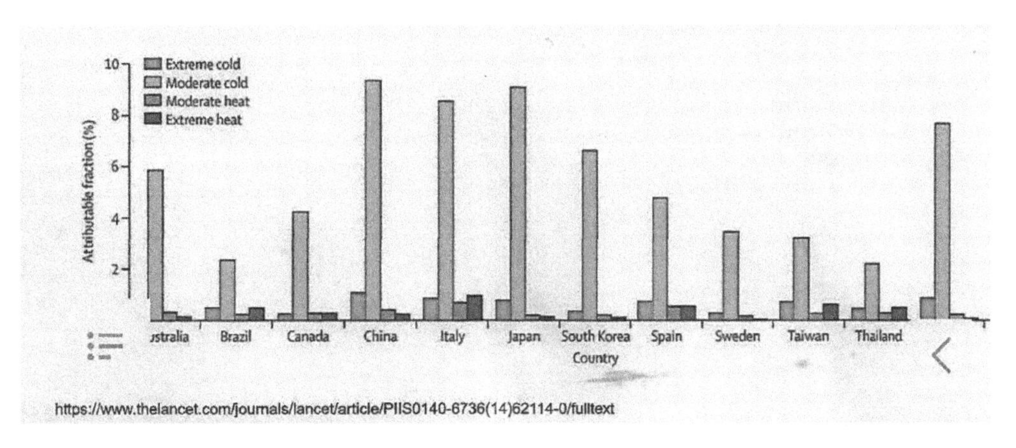

[177] NOAA National Hurricane Center statistics. Cited by Prof Roy Spencer 26th August 2020 wattsupwiththat.com.

[178] See Pauli et al 1996 *"Effects of climate change on mountain eco-systems"* World Resources Review" 8. 382 – 390.

[179] Professor Plimer *Heaven and Earth* Op cit p195.

[180] Professor Easterbrook Op Cit p 115.

In 2003[181] a study was made of annual excess mortality in the USA from 1964 to 1998 on days when 'apparent' temperatures (combining air temperature and humidity) exceeded a threshold value for 28 major cities. For the 28-city average, there were 41.0 excess heat-related deaths per year per million in the 1960s and 1970s, 17.3 in the 1980s and 10.5 in the 1990s. Cities such as Tampa Florida and Phoenix Arizona, with high incidence of heat waves, had the lowest heat related mortality.

The report concluded that *' this systematic desensitization of the metropolitan populace to high heat and humidity over time can be attributed to a suite of technologic, infrastructural, and biophysical adaptations, including increased availability of air conditioning'.*

On a global scale an analysis of 75,225,200 deaths in 13 countries over the period 1985 – 2012 with wide ranges of climate conditions revealed that cold weather kills 20 times more than hot weather.

Moreover, there are highly researched studies that show deaths attributable to extreme weather conditions occur predominantly in winter months as illustrated by the following graph.

Figure 46

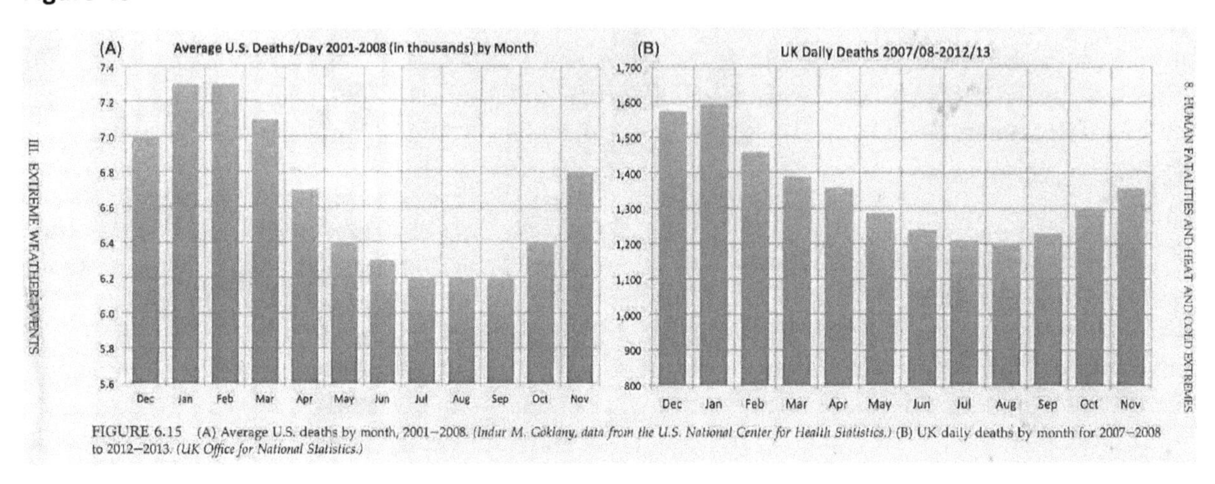

FIGURE 6.15 (A) Average U.S. deaths by month, 2001–2008. *(Indur M. Goklany, data from the U.S. National Center for Health Statistics.)* (B) UK daily deaths by month for 2007–2008 to 2012–2013. *(UK Office for National Statistics.)*

SUMMARY

Extreme weather events, changes in sea levels, coral growth, extinctions and heat related deaths are not indicators of changes in climate much less global warming induced by human emissions.

[181] Davis, Knappenberger, Michaels, and Novicoff, 2003.

CONCLUSION

The 2015 Paris conference agreed that action was needed to keep global temperature increase desirably to below 2^0C above pre-industrial levels and essentially below 1.5^0C . Temperature has already increased since 1890 by over 1^0C. Moreover the Medieval Warming Period and the Roman Warming Period enjoyed temperatures higher than those of today and were times of great agricultural yields, prosperity, and cultural advances.

The UK is responsible for just 1.4% of the global total of CO_2 emissions – less than China adds to the atmosphere each year. Very few nations have legislated to cut CO_2 emissions in line with the limits proposed at the Paris conference. In 2020 China built more than three times as much new coal power capacity as all other countries in the world combined. India, Indonesia, and Vietnam are rapidly building yet more coal fired power stations.

In 2015 two thirds of our electricity in the UK was generated by fossil fuel power plants.

We have now shut down most of our coal fired plants.

Given all this is it not astonishing to recite that the UK government has legislated to eliminate net emissions from fossil fuels by 2050 and to reduce them by 78% from 1990 levels by 2035.

Already industry and commerce are suffering due to the rocketing price of electricity compared to our trade competitors abroad. On Friday 3rd September 2021 electricity prices reached an all-time high of £240 per megawatt hour. On Monday 6th September 2021 the UK was getting 45.6% of its power from gas fired turbines, 15.5% from nuclear power plants, 12.3% from interconnectors to the Continent and Northern Ireland and 5.55% from coal. Wind output had dropped to 474 megawatt from 14,286 megawatts in May. National Grid EST asked Électricité de France to switch on two coal fired units at its West Burton A station to help meet demand. Soaring world gas prices has led to a resurgence of coal fired plants[182].

The statutory UK body, the Climate Change Committee, has declared that there is very little likelihood of achieving these targets. To do so would involve massive de-carbonisation of industry and power generation as well as of all forms of transport. Electricity to sustain our economy cannot possibly be met by solar and wind generation[183].

Sudden, disruptive and immediate action is inevitable if these levels are to be achieved. The staggering cost will fall on all of us.

All to achieve a negligible percentage fall in atmospheric CO_2.

All for a hypothesis which is false.

[182] *Daily Telegraph* 7th September 2021 Business.

[183] "We can't install enough wind to meet our needs. We're going to need something else" Andrew Crossland Fellow of Durham Energy Institute reported in *Daily Telegraph* 7th September 2021.

www.brugesgroup.com

DONATE TO THE BRUGES GROUP

Yes, I wish to donate to *The Bruges Group*

☐ £5 ☐ £10 ☐ £20 ☐ £50 ☐ £100 ☐ £250 ☐ £500 ☐ £1,0[illegible]

Other please specify:.......................................

DONATIONS BY CHEQUE / POSTAL ORDER

Title: .... First Name:........................

Surname: ....

Address: ....

.... Postcode ....

Email:........................

Telephone: ....

Signature: .... Date:........................

PLEASE MAKE CHEQUES PAYABLE TO ***THE BRUGES GROUP***

PLEASE RETURN THIS FORM TOGETHER WITH YOUR DONATION TO:

The Bruges Group, 246 Linen Hall, 162-168 Regent Street, London W1B 5TB

DONATIONS BY CREDIT / DEBIT CARDS

☐ MasterCard ☐ VISA ☐ ☐ JCB ☐ MasterPass ☐ Maestro

Card number ☐

Valid from ☐ Expiry date ☐ Issue number ☐ Security code ☐

Card holder's name as it appears on the card (please print):........................

....

Address of card holder:........................

.... Postcode ....

Email:........................

Telephone: ....

Signature: .... Date:........................

OR AT ANY BARCLAYS BANK

Account Name: The Bruges Group **Sort Code:** 20-46-73 **Account number:** 90211214

FOR DONATIONS OVER THE PHONE CALL OUR HOTLINE ON 020 7287 4414

ONLINE DONATIONS

Secure online donations can be made via our website at **www.brugesgroup.com/donate**
Donations can be made via credit / debit cards

DONATIONS BY STANDING ORDER

Yes, I wish to donate to *The Bruges Group*

Title: .... First Name: ....

Surname: ....

Address: ....

....

.... Postcode ....

Email: ....

Telephone: ....

Please complete this form in BLOCK CAPITALS and return it to us at the address overleaf.

To: The Manager, ... Bank/Building Society

Branch Address: ....

....

.... Postcode: ....

Your Account Number ☐☐☐☐☐☐☐☐ Sort code ☐☐ ☐☐ ☐☐

Please pay Barclays Bank PLC (sort code 20-46-73) 6 Clarence Street, Kingston-upon-Thames, Surrey KT1 1NY

The sum of (please tick as appropriate) ☐ £10 ☐ £20 ☐ £50 ☐ £100 ☐ £250 ☐ £500 ☐ £1,000

☐ Other, please specify: ....

.... Amount in words

To the credit of *The Bruges Group*, Account Number 90211214
forthwith and on the same day in each subsequent **year** or **month** *(please circle your preference)* until further notice.

Signature: .... Date: ....

Please print name and title: ....

THE BRUGES GROUP | ASSOCIATE MEMBERSHIP

TEL: +44 (0)20 7287 4414 | JOIN NOW | info@brugesgroup.com

To become an Associate Member of the Bruges Group, complete the following form and send it to the Membership Secretary with your annual subscription fee. This will entitle you to receive our published material for one year. It also helps cover the cost of the numerous Bruges Group meetings to which all Associate Members are invited. **You can also join online, right now, by using your debit or credit card. Please log on to www.brugesgroup.com/contact-us/join-now or you can join over the phone by calling 020 7287 4414.**

Minimum Associate Membership Rates for 1 year UK Member £30 ☐, Europe £45 ☐, Rest of the world £60 ☐

Lifetime Membership: £500 ☐ Optional donation: £10 ☐ £20 ☐ £50 ☐ £100 ☐ £250 ☐ £500 ☐

Other, please specify: ....

If you are able to give more towards our work, we would be very grateful for your support. For the sake of convenience, we urge you to pay by standing order.

YES! I wish to become an Associate Member of the Bruges Group

Title: .... Name: ....

Address: ....

.... Postcode: ....

Telephone: ....

Email: ....

BANKERS ORDER Name and full postal address of your Bank or Building Society

To: The Manager:.......... Bank/Building Society

Address: ....

.... Postcode: ....

Account number: .... Sort code: ....

Please Pay: Barclays Bank Ltd (Sort Code 20-46-73), 6 Clarence St, Kingston-upon-Thames, Surrey KT1 1NY

The sum of £ .... (figures)

Signature: .... Date: ....

to the credit of the Bruges Group A/C No 90211214 forthwith and on the same day in each subsequent year until further notice.

— or —

CHEQUE PAYMENTS I enclose a cheque made payable to the Bruges Group

The sum of £ .... (figures)

Signature: .... Date: ....

— or —

MEMBERSHIP PAYMENT BY CREDIT/DEBIT CARD

☐ MasterCard ☐ ☐ VISA ☐ JCB ☐ Maestro ☐ MasterPass

Card number: ....

Valid from: .... Expiry date: .... Issue number: .... Security code: ....

Card holder's name as it appears on the card (please print): ....

.Address of card holder: ....

.... Postcode: ....

Telephone:..........

Email: .... Signature: .... Date: ....

www.brugesgroup.com

Please complete this form and return to:
The Membership Secretary, The Bruges Group, 246 Linen Hall, 162-168 Regent Street, London W1B 5TB

THE BRUGES GROUP

The Bruges Group is an independent all-party think tank. Set up in 1989, its founding purpose was to resist the encroachments of the European Union on our democratic self-government. The Bruges Group spearheaded the intellectual battle to win a vote to leave the European Union and against the emergence of a centralised EU state. With personal freedom at its core, its formation was inspired by the speech of Margaret Thatcher in Bruges in September 1988 where the Prime Minister stated, "We have not successfully rolled back the frontiers of the State in Britain only to see them re-imposed at a European level."

We now face a more insidious and profound challenge to our liberties – the rising tide of intolerance. The Bruges Group challenges false and damaging orthodoxies that suppress debate and incite enmity. It will continue to direct Britain's role in the world, act as a voice for the Union, and promote our historic liberty, democracy, transparency, and rights. It spearheads the resistance to attacks on free speech and provides a voice for those who value our freedoms and way of life.

WHO WE ARE

Founder President: The Rt Hon. the Baroness Thatcher of Kesteven LG, OM, FRS
Vice-President: The Rt Hon. the Lord Lamont of Lerwick
Chairman: Barry Legg
Director: Robert Oulds MA, FRSA
Washington D.C. Representative: John O'Sullivan CBE,
Founder Chairman:
Lord Harris of High Cross
Former Chairmen:
Dr Brian Hindley,
Dr Martin Holmes
& Professor Kenneth Minogue

Academic Advisory Council:
Professor Tim Congdon
Dr Richard Howarth
Professor Patrick Minford
Ruth Lea
Andrew Roberts
Martin Howe, QC
John O'Sullivan, CBE

Sponsors and Patrons:
E P Gardner
Dryden Gilling-Smith
Lord Kalms
David Caldow
Andrew Cook
Lord Howard
Brian Kingham
Lord Pearson of Rannoch
Eddie Addison
Ian Butler
Thomas Griffin
Lord Young of Graffham
Michael Fisher
Oliver Marriott
Hon. Sir Rocco Forte
Graham Hale
W J Edwards
Michael Freeman
Richard E.L. Smith

BRUGES GROUP MEETINGS

The Bruges Group holds regular high–profile public meetings, seminars, debates and conferences. These enable influential speakers to contribute to the European debate. Speakers are selected purely by the contribution they can make to enhance the debate.

For further information about the Bruges Group, to attend our meetings, or join and receive our publications, please see the membership form at the end of this paper. Alternatively, you can visit our website www.brugesgroup.com or contact us at info@brugesgroup.com.

Contact us

For more information about the Bruges Group please contact:
Robert Oulds, Director
The Bruges Group, 246 Linen Hall, 162-168 Regent Street, London W1B 5TB
Tel: +44 (0)20 7287 4414 **Email:** info@brugesgroup.com

www.brugesgroup.com

Lightning Source UK Ltd.
Milton Keynes UK
UKHW051440201121
394201UK00004B/62